RSHORE GROUP O
ARNING RESOURC   KV-571-512

Grower Talks®
on
Crop Culture 2

Edited by
Rick Blanchette and Jayne N. La Scola

PERSHORE & HINDLIP
COLLEGE LIBRARY

4/00

CLASS	CODE
635.9	021787

WITHDRAWN

Ball Publishing
Batavia, Illinois
USA

021787

LIBRARY
PERSHORE COLLEGE
AVONBANK
PERSHIRE
...RSHIRE WR10 3JP

Ball Publishing
335 North River Street
Batavia, IL 60510 USA
www.ballpublishing.com

Copyright © 1999 Ball Publishing. All rights reserved.

Photos on pages 110, 114, and 117 are courtesy of the Netherlands Flower Bulb Information Center.

Cover designed by Tamra Bell.

Cover photo of lavender crop © Bryan Peterson/AGStockUSA.
All rights reserved.

No part of this book may be reproduced or transmitted in any form by any means, electronic or mechanical, including photocopying, recording, or any other storage and retrieval system, without permission in writing from Ball Publishing.

Reference in this publication to a trademark, proprietary product, or company name is intended for explicit description only and does not imply approval or recommendation to the exclusion of others that may be suitable.

Library of Congress Cataloging-in-Publication Data

GrowerTalks on crop culture 2 / edited by Rick Blanchette and Jayne N. La Scola.
    p. cm.
  ISBN 1-883052-21-1 (alk. paper)
  1. Floriculture. 2. Plants. Ornamental. I. Blanchette, Rick, 1966- .
II. La Scola, Jayne N., 1975- . III. GrowerTalks. IV. Title: GrowerTalks on crop culture two.
SB405.G79   1999
635.9—dc21

                            99-28399
                              CIP

Printed in the United States of America
04 03 02 01 00 99 1 2 3 4 5 6

Contents

Primula

Protea

Roses

Rudbeckia

Schizanthus

Schlumbergera (Holiday Cacti)

Shamrock

Snapdragon

Spathiphyllum

Sunflower (Helianthus)

Trachelium

Introduction

For over sixty years, *GrowerTalks* has been the authoritative voice in floriculture. Growers have long trusted the information on industry trends, marketing, propagation, and cultivation. Here in one volume are over fifty articles from the pages of *GrowerTalks'* Culture Notes department.

Written by industry experts, the articles provide tips for the cultivation of over forty crops from aglaonema to trachelium. Poinsettias, chrysanthemums, geraniums, herbs, perennials, and lilies all receive special attention. We've even included tips on producing and marketing hanging baskets.

You'll learn proven propagation techniques and how to get the most from your crop. Each chapter includes valuable information about planting, watering, nutrition, pest management—even marketing and presentation.

With *GrowerTalks on Crop Culture 2*, you'll gain experience *before* you need it.

Sixty Years of Culture and Crops

Joli A. Shaw

In the six decades since *GrowerTalks'* first issue in May 1937, floriculture production has undergone amazing cultural revolutions, from gravel culture in the '30s and '40s to mum lighting in the '50s to DIF in the '80s to fourteen-day bedding in the '90s. We've pulled together some of the most interesting and innovative cultural developments to showcase for you.

We've also taken a look at some of the popular crops of the eras. Though their uses have changed and their popularity has fluctuated over time, these crops have survived market whims and preferences to keep their places as standards in production.

Here, in *GrowerTalks'* own words, are some of the highlights of culture and crop trends through the decades.

GrowerTalks' Top Picks for Cultural Breakthroughs

Gravel culture

First published in *GrowerTalks* in 1938, gravel culture for cut flowers was appealing for its "elimination of all soil-borne troubles such as soil diseases and insects, fertilizers, deficiencies or excesses, drainage and aeration troubles." Growers were advised to treat gravel with formaldehyde before planting to prevent rot. But after several years of experimentation and scattered commercial use, gravel culture faded away, in part due to the development of soilless mixes.

Soilless mixes

The famous UC mixes: developed at the University of California, Los Angeles, some of the first soilless mixes were peat and sand and were introduced in the mid-1950s. They were made up of fine sand "of the wind blown or water deposited types" and European or Canadian peat moss. Pot plant growers were advised to use a 50/50 mix, bedding plant growers a 75% sand/25% peat mix. The mixes had very low fertility, and growers had to add nitrogen, phosphate, calcium, magnesium, and other nutrients.

Tissue culture

Propagation was immeasurably speeded with the development and application of tissue culture, which *Grower Talks* first reported on at Yoder Brothers on carnations in the early '60s. However, it wasn't used extensively in our industry until the '80s, when researchers began experimenting with foliage plants, gerbera, ferns, violets, orchids, bromeliads, and others. By 1983, a Florida tissue culture lab was shipping out 1.25 million plantlets per year to growers all over the U.S.

DIF

Pioneered at Michigan State University, DIF was first used on Easter lilies in the late '80s. The idea of using warm night temperatures and cool day temperatures revolutionized chemical-free height control. Growers and researchers soon expanded the practice to control height on poinsettias and a variety of bedding and pot plants.

Fourteen-day bedding

When Vic was writing *Grower Talks* in late 1961, the biggest news in bedding plant production was eight-week petunias. Now in the '90s, fourteen- and even ten-day bedding crop times from plugs have enabled growers to get an unprecedented three or four turns out of their ranges, increasing both production and profitability.

Grower Talks' Favorite Crops through the Years

Seed geranium

The first successful seed geranium series, Carefree, featured in *Grower Talks'* August 1967 issue, began America's love affair with seed geraniums. Still arguably a top bedding plant today, seed geraniums have expanded beyond bedding and are now used in pots, hanging baskets, patio pots, color bowls and window boxes.

Pot chrysanthemum

Though cut mums were always a staple item for growers, putting mums in pots proved to be the real sales push for this crop. Their real popularity came in the late '50s and '60s when the mum market exploded with new varieties and colors. The April 1957 *Grower Talks* reported that the wholesale value of mums had approximately doubled since the 1949 census. Growers were looking for new mum varieties, earlier varieties, and searching for new techniques, even experimenting for a time with flashing light (two seconds out of every minute) instead of continuous light. New growth regulating chemicals made them even more suitable as home pot plants. But it wasn't until the development of year-round mum production that pot mums could break through the barrier of being only a fall crop and get into other markets such as Mother's Day, ultimately forming the platform for weekly flowering pot plant sales in supermarkets.

Pansy

An important bedding plant even in the '40s and '50s, pansies have experienced a recent resurgence in popularity as growers found a new fall market for the crop, even competing with garden mums. Growers liked them for their ease of growing and insect and disease resistance, *GrowerTalks* reported in August 1949. Growers even put them in pots in the '50s and sold them for home use. They've gone from mainly being produced in open beds to exclusively being grown in greenhouses. F_1 hybrids have propelled pansies into the bedding plant top five.

Petunia

Always a popular bedding plant choice, petunias, especially the double varieties, were discussed more often as a pot plant in the early *GrowerTalks*. But the petunia market was forever changed with the introduction of the F_1 hybrid singles, though they were slow to gain popularity. This was "partly due to the well entrenched buying habits of florists, and certainly partly due to their price," according to the December 1951 *GrowerTalks*. Still, we recommended them for "customers looking for the best regardless of price."

Snapdragons

Starting as a staple cut crop, then moving into a popular bedding plant, snapdragons have long been part of growers' programs. The cover of the May 1940 *GrowerTalks* declared, "Better strains of standard varieties and new colors in forcing snapdragons are steadily increasing their value for cutting." A guide to snapdragon culture in the February 1950 issue of *GrowerTalks* advised growers, "Under the high light and temperature conditions of the cloth house, we don't get the fine 3-foot stems that you see under glass in February." It wasn't until June 1951 that *GrowerTalks* reported on single stem snap culture, telling growers the method produced a higher percentage of long grade A spikes than with a pinched crop.

Joli A. Shaw is a freelance writer and former associate editor of GrowerTalks *magazine, Lisle, Illinois. May 1997.*

Aglaonema

Fine-tuning Your Aglaonema Production

Lynn Griffith

Aglaonemas are produced commercially in most parts of the world as freestanding specimens. Pot sizes generally range from 4 to 10 inches, though other sizes are occasionally produced. Most of the time three to six cuttings per pot are used, usually placed close together in the container. Three 6-in. pots are sometimes placed in 14- or 17-in. pots for large interiorscape specimen production. Cuttings are sometimes included in dish gardens. Hanging basket production is rather rare. Interiorscape plantings frequently use 6- to 10-in. plants. Because of cold sensitivity, landscape plantings of aglaonemas are limited to tropical areas. They excel in interiors because of tolerance of low-light conditions.

Varieties

Silver Queen is probably the most widely produced aglaonema cultivar in the world. Plants have grayish-green leaves with white and silver to gray blotches. Silver Queen suckers fairly well, though it's rather cold sensitive (58° F/14° C) and susceptible to bent tip. As winter progresses, plants become hardier and may tolerate 45° F (7° C).

Maria (also called Emerald Beauty) is more compact and darker than Silver Queen, with deep green foliage highlighted by bands and flecks of silvery green. Maria grows a little more slowly than Silver Queen and is rarely seen in larger than 10-in. pots. Emerald Beauty is more cold-tolerant (45° F/7° C).

Propagation

Most commercial propagation is from cuttings. Callused and rooted cuttings develop a little faster and are sometimes less disease prone, although they cost more. Cuttings should generally have five leaves. Cuttings will sucker better if they're spaced in the pot as opposed to being potted in a clump, though both methods yield a marketable plant. Rooting time is about four weeks (one week in warm soil, longer in cooler soil). Growers occasionally mist cuttings during the warm part of the day, but excessive misting can lead to disease problems. Rooting hormones produce inconsistent results, and many growers don't use them.

Culture

Various combinations of peat, bark, wood chips, sawdust, sand and perlite are used successfully. As long as the mix has decent aeration and moisture-holding capacity, aglaonemas tend to grow pretty well. The target is usually 5.5 to 6.5

pH, though they can tolerate lower pH. Plants favor moist conditions and are fairly heavy feeders, though not particularly fast growers. Aglaonemas frequently are drenched shortly after potting with a fungicide combination containing thiophanate methyl (Clearys 3336, Domain, etc.) plus Subdue, Aliette, or Truban for water mold fungi. Banrot is also registered. High-phosphate, soluble starter fertilizers tend to help expand root systems once cuttings have rooted. Six-in. pots with three to five cuttings per pot are generally grown on 8- to 10-in. centers, usually on ground cover, sometimes on a bench. It normally takes six to eight months to finish a 6-in. Silver Queen, depending on cutting size and the number of cuttings used. Maria tends to take a little longer. Best light levels are 1,500 to 2,500 foot-candles, or about 80% shade in the tropics. When light is too high, foliage stays rather vertical with pale color and tan blotches near leaf tips. When light is too low, plants look good but grow very slowly.

Nutrition

Most aglaonema varieties are relatively heavy feeders, requiring about 1,200 lbs. of nitrogen per acre per year or 1,400 lbs. in shadehouses exposed to rainfall. They have somewhat high requirements for potassium, magnesium, iron, and copper. Aglaonemas can be grown:

- with granular fertilizer as a top-dress;
- with coated slow-release fertilizers applied either as a top-dress or incorporated into the soil mix;
- with constant liquid or soluble fertilizers; or
- with a combination of these methods.

They do well with a 3-1-2 ratio of nitrogen-phosphate-potash, frequently with supplemental magnesium.

Nitrogen deficiency gives you small, pale leaves and little growth. Phosphorus deficiency is rather rare, usually resulting in a weak root system. Potassium-deficient plants show necrosis in the older foliage, and plants tend to shed older leaves. Magnesium deficiency is common, with broad yellowing of margins of older leaves. It's especially common in darker varieties that have more chlorophyll. Plants low in calcium have a thinner, softer leaf blade and weak foliage overall. Lack of iron shows typical interveinal chlorosis in new foliage. Copper deficiency is quite rare today because of the use of copper fungicides and evolution toward new varieties. Lack of copper gives you a crippling of the new leaf, resulting in a small, barely formed leaf. Boron toxicity is manifested as tan to brown blotches just back from the leaf tip on the larger half of the leaf blade.

Diseases

The most serious disease of aglaonemas is erwinia bacteria, both *Erwinia carotovora* and *E. chrysanthemi*. Erwinia causes bacterial soft rot of stem or leaf tissue. Symptoms are a wet, slimy rot of leaves or stems, sometimes but not always associated with a foul odor. Erwinia problems are common during propagation and somewhat less so after plants are rooted. Erwinia stem rot is a wet, mushy rot that turns foliage yellow. The condition is best prevented with cultural controls and the use of clean cuttings and a clean greenhouse. Growers commonly use copper fungicides frequently mixed with Dithane.

Fusarium stem rot is another common disease that frequently originates in stock plants. It's a drier rot than erwinia, with the internal stem tissue appearing white and somewhat mealy inside. Edges of infected stem tissue frequently have a purple or red appearance. Thiophanate methyl fungicides are generally used against fusarium stem rot, sometimes in combination with Captan. Keeping soil pH up may help as well. Pythium, another common root rot fungus of aglaonema, generally occurs under wet conditions or in heavy, poorly drained soils. Subdue, Truban, and Aliette combat pythium.

Nematodes, specifically lesion (*Pratylenchus* sp.) and root knot (*Meloidogyne* sp.) are a major problem in aglaonemas. Roots become hollow—like a drinking straw—so only the root cortex remains. Plants lose vigor and leaf size, and cuttings droop. Plants also try to send out new roots higher up on the stem. Stock plant growers generally treat two to three times a year with nematicide, using such items as Counter, Puradan, Mocap, Temik, and Vydate, though it's unlikely that any of these are registered for their situation.

Oxamyl is a possible control measure in nurseries. Nematodes can come in on uprooted or rooted cuttings, and once the population builds up, plants decline later in life. Hot-water cutting dips at 122° F (50° C) deserve further investigation.

The most common leaf spots of aglaonema are myrothecium and collectotrichum. Myrothecium frequently attacks wounded tissue, especially in propagation. It's a large, dry, brown leaf spot, usually circular with visible white and black fruiting bodies on the foliage undersides. Removing affected leaves is helpful, and sprays of Dithane or Daconil are common. Colletotricium is similar to myrothecium, but it tends to spread, becoming more of a leaf blight. Sprays are similar to myrothecium, though copper fungicides are also used.

Two less common diseases are xanthomonas and dasheen mosaic virus. Plants infected with xanthomonas show brownish and yellowing tissue along margins and leaf tips, especially in older leaves. Symptoms look more like a burn or toxicity rather than a disease. Lesions aren't water soaked, and no fruiting bodies are found. Bacteria spreads primarily by splashing water, and the most effective control is to use sterilized tools to remove infected leaves. Copper and dithane sprays are also used, as are Phyton 27 or Aliette. Dasheen mosaic virus is rare today. It causes a mosaic or blotching symptom on the foliage, and plants tend to be stunted with some leaf distortion. It spreads primarily by cutting tools, though aphids can spread it. Don't waste your time with virus-infected stock—throw it away, sterilize, and start again from a different cutting source.

Insects

Foliar and root mealybugs are the most common insect pests. They're small, white-segmented insects, frequently with red juice inside them. Growers usually use Diazinon, Spectracide, or Dycard for foliar mealybugs. Talstar and Cygon are also frequently used. Root mealybugs are usually controlled with Diazinon drenches.

Problems with scales, aphids, and mites are rare in aglaonemas. Thrips can be a problem at times, working within the tube of the unfurled leaf. Leaves tend to be torn on one side and have numerous small, brown blotches similar to a disease symptom. Mavrik, Orthene or Avid sprays are common, especially with a wetting agent or penetrant. Several sprays used in rotation at three- to four-day intervals are usually necessary.

An occasional problem that looks like the work of an insect but isn't is the birdsnest fungus, also known as shotgun fungus or glebal masses. The fungus looks like a scale insect, with small, brown disks on undersides. The fungus is actually soil-borne, but fruiting bodies of the fungus show up on foliage. Control the fungus by removing affected leaves and using clean soil. Captan sprays seem to help, too.

Disorders

Bent tip is a common aglaonema disorder affecting Silver Queen primarily, as well as a few other varieties. It doesn't have a nutritional cause. High light levels and water stress are often blamed. However, it also seems that when aglaonemas grow too fast, the tip of the leaf catches itself early in the unfurling process, creating a bent tip symptom. The disorder isn't a big production problem, and it's fairly rare under interior or homeowner conditions where growth is slower. Tip burn is a common symptom when soluble salts are excessive. A dry tan or brown symmetrical tip results.

Cold injury is rather distinctive on aglaonemas, usually resulting in a dark, greasy appearance on the upper leaf surface. It's the result of epidermal collapse and tends to affect older leaves as well as older plants.

Cultural Tricks

Leaf size and thickness can be increased significantly with calcium nitrate sprays using about 2½ lbs. per 100 gal. of chelated calcium. Most chelated iron products work as a drench in controlling iron deficiency in this plant, but only the EDDPA chelates such as Sequestrene 138 work as a foliar spray.

Aglaonema cuttings can be rooted in water, and this practice is commercially viable. In Puerto Rico, for example, some growers root plants in plastic cups filled with ½ in. of water. In Central America, plants are rooted by placing cuttings in sheaths of black plastic. Benzyladenine sprays will cause aglaonemas to sucker profusely, although this hasn't been commercially proven.

Bonzi has been used to control height in large, stretchy varieties such as Abidjan. Aglaonema roots don't like light, and they tend to grow better root systems in black or green pots, rather than opaque white ones. Roots tend to be more in the integral part of media when growing in opaque pots. Sprays of 250 ppm gibberellic acid are used to induce flowering for breeding purposes.

Lynn Griffith is president of A & L Southern Agricultural Laboratories, Pompano Beach, Florida, and is the author of Tropical Foliage Plants: A Grower's Guide. August 1996.

Alstroemeria

Alstroemeria

Dr. Mark P. Bridgen

Alstroemeria, also called Peruvian lily, Lily of the Incas, or Inca lily, has been grown in the United States since the 1970s as a cut flower crop, but now is also being grown as a garden flower and as a potted plant.

The Inca lily's popularity is increasing in the floriculture industry because it's versatile and easy to grow under cool temperatures. Flower colors include red, pink, purple, lavender, white, yellow, orange, and bicolors; each flower lasts three to four weeks on the plant and survives ten to fourteen days as a cut flower. Flowers may be harvested anytime during the flowering season since the underground rhizomes will continue to produce flowering stems.

Alstroemeria is perennial in the South and blooms from February to June and again from September until November. Along the West Coast, it flowers all summer. If alstroemeria is planted at pansy time in temperate regions, it will flower from June until frost, except in hot, dry conditions. Alstroemeria is considered an annual above Zone 6, though its degree of winter hardiness depends upon where the hybrid was developed, the climate in which it's grown, and the type of overwintering protection provided.

Propagation

Alstroemeria seed sources are few and availability varies. To hasten fresh seed germination, pour boiling water over seeds and soak them for eight hours or overnight. Replace the water at least once during soaking for best results. Plant seeds in a well-drained medium, keep them moist, and grow at 75° F (24° C). Old seeds may not germinate readily; if the seeds haven't germinated after four weeks, refrigerate them for four weeks or until they begin to germinate. After the cold treatment, move the seedlings and any ungerminated seeds to 75° F (24° C) for four to eight weeks.

Alstroemeria is usually vegetatively propagated by rhizome division or micro-propagation to obtain true-to-type plants that reach productive size faster than seed-propagated plants. Alstroemeria produces underground rhizomes with thick storage roots that provide an immediate carbohydrate supply for the growth of asexually propagated plants. However, the rhizomes' storage roots take up more pot space than seedlings and can be difficult to package for sale.

Transplants

If you buy alstroemeria as 2½-in. liners, plant them directly into the final pot. Any well-drained medium containing ample organic matter and 6 to 7 pH can be used. Shallow planting produces earlier flowering and bushier plants.

To produce container plants for spring sales, pot alstroemeria in the fall. Once liners are established, temperatures can be dropped as low as 38° F (3° C), and roots will continue to grow. About ninety to one hundred days prior to sale, cut the foliage back completely, and move the pots to warmer temperatures.

Watering

Proper watering is the key to success with alstroemeria. Rhizomes of newly established plants will quickly rot if overwatered, but established plants need abundant water. If they dry out for a few days in extreme heat, the foliage will turn yellow and flower buds will abort. Plants will induce new shoots when they are watered again, but a drought period will delay blooming for several weeks.

Light

Alstroemeria grows best in full sun with a mulched soil surface and adequate water. Plants grown in the shade produce fewer flowers, though morning sun is often adequate. Long days, produced by low-intensity incandescent lighting or HID lighting, hastens flower initiation. A thirteen-hour minimum of light per day is recommended for most cultivars.

Fertilization

Once established, alstroemeria plants require high nutrient levels. Research shows that the flower number increases linearly by increasing nitrogen to 400 ppm each week.

Growing Temperatures

Greenhouse night temperatures of 50° to 60° F (10° to 16° C) and day temperatures of 65° to 70° F (18° to 21° C) are optimal. Prolonged temperatures above 75° F (24° C) may decrease or stop flowering. However, rhizome temperatures are more important than air temperature and should be kept at 55° to 60° F (13° to 16° C) to induce flowering. Late afternoon waterings and mulching help keep rhizomes cool during hot periods.

Height Control

No commercial growth regulator is registered for controlling alstroemeria height. Keep plants short by choosing compact cultivars and by increasing light intensity with adequate spacing. Final spacing on 15-in. centers for 7-in. pots and 18-in. centers for 8-in. pots produces compact plants.

Alstroemeria demonstrates a physiological effect called the short stem effect. The more the plants are pruned, the less they will grow, so it's beneficial to remove unsightly and irregular stems periodically. Also, plants that are grown in cooler temperatures will have shorter, thicker stems than plants grown in warmer temperatures.

Pests and Disease

Inca lilies are relatively disease- and pest-free, though snails, slugs, aphids, caterpillars, and whitefly can cause problems. Thrips are the most devastating pests because they get inside the flowers, are difficult to control, and may transmit deadly viruses.

Botrytis and root rots may appear during low light periods. Avoid botrytis by maintaining good air circulation, removing infected plant parts, and using preventative fungicides. Fungicide drenching of newly planted pots is recommended.

Availability

American-bred alstroemeria liners are available wholesale from Mojonnier Enterprises in Encinitas, California; Coast Alpine Nursery in Lummi Island, Washington; and from the University of Connecticut, Storrs, Connecticut. Dutch varieties are available from the Fred C. Gloeckner Co. of New York, New York.

Dr. Mark Bridgen is associate professor of floriculture, Department of Plant Science, University of Connecticut. November 1994.

Amaryllis (Hippeastrum)

Forcing Pot Amaryllis: Keys to Success

A. A. De Hertogh

Amaryllis (hippeastrum) originated in South America. Flowering is regulated by bulb size, temperature, and moisture. The primary sources of amaryllis bulbs forced in the U.S. and Canada are Israel, South Africa, and the Netherlands.

The major use is for pot plant forcings, but they also can be used as cut flowers. The general marketing season is from September to May. Normally, the South African–grown cultivars are forced early, the Israeli- and Dutch-grown cultivars medium to late. With special growing and handling, though, some Israeli and Dutch cultivars are suitable for December forcings. The objective is to market a plant that has simultaneously produced growing leaves and at least one floral stalk.

The number of floral stalks produced is influenced by bulb size and cultivar. Examples of commercial-sized bulbs are (in circumference) 20/22, 24/26, 28/30 and 32+ cm. The number of flowers per stalk is primarily a cultivar response, but most cultivars produce four flowers per stalk. The range is two to six. Larger bulbs tend to produce two floral stalks.

Planting

After harvest, bulbs are quickly dried and cured. During this and all subsequent processes, it's critical that the old root system be kept viable. Normally, bulbs are cured for two weeks at 73° to 77° F (23° to 25° C) with high ventilation rates. They're subsequently stored at 48° to 55° F (9° to 13° C) at 80% relative humidity for at least eight to ten weeks. Hold bulbs stored for longer periods at 41° to 48° F (5° to 9° C). Transport bulbs at 48° F (9° C). In addition, bulbs must be protected against freezing and drying out.

Forcers should be prepared to plant bulbs as soon as they arrive. If they must be stored, place them at 41° to 48° F (5° to 9° C). The precise temperature for preplant storage will depend on the sprouting condition of the bulbs upon arrival. If they have begun to sprout, store them at 41° F (5° C). If no sprouting is observed, store them at 48° F (9° C). Keep bulbs from drying out during preplanting storage.

Plant amaryllis in well-drained, sterilized planting media with pH of 6.0 to 6.5. Never use fresh manure or bark as part of the media. Growing media must be capable of being packed tightly around the roots.

Normally, one bulb is planted per 6-in. standard pot. Plant the bulb with the nose above the rim of the pot; one-third of the bulb should be above the planting medium. Force bulbs pot-to-pot on the bench.

After planting, water media thoroughly. Keep media only slightly moist. It's important not to overwater the plant in order to stimulate regrowth of the basal root system. Normally, watering once per week is satisfactory. Use tepid water, and don't water over bulb noses.

Initially, bulbs don't need fertilization. After they're marketed, however, consumers should be advised to fertilize plants.

The primary disease of amaryllis is fire or red spot (stagnospora). Overwatering may promote the development of fusarium. In addition, it's possible to have mites, thrips, and mealybugs.

Forcing

Amaryllis is a tropical plant, and you can force it over a wide range of temperatures, but 70° to 80° F (21° to 27° C) is preferred (see table). Use bottom heat.

Force plants in a greenhouse with medium light intensity (2,500 to 5,000 f.c.). You can start bulbs in a dark, temperature-controlled area before you place them under lighted conditions. Force plants in a well-ventilated greenhouse. Don't let relative humidity build up.

The average forcing time to the market stage of development is three to five weeks. It will vary with each cultivar and forcing period (see table). It's also important to note that most lots are variable. Thus, the forcing information in the table should be used only as a guide to average dates of marketing and flowering.

Marketing

Market plants when the floral stalks are 12 in. tall. At marketing, it's desirable to have leaf growth of 6 to 12 in. and a second stalk beginning to grow. Don't cold-store plants. If you have to hold them, place them at 48° F (9° C). Wholesalers and retailers should use tepid water after they receive plants.

Whenever possible, market plants with care tags. Consumers should be informed that amaryllis needs to be fertilized at least one or two times per month when it's growing. They should keep plants in the coolest area of the home and out of direct sunlight in order to obtain maximum life from flowers. Amaryllis can be placed outside in pots when the danger of frost has passed.

To reforce plants, two systems are available. You can take them into the home in the fall, allow them to dry and store them for at least eight weeks at 50° to 60° F (10° to 16° C). Then, cut off dried leaves, water the planting media, and place plants in a warm area to start the forcing process. If you don't want to store the bulbs, you can grow the plants in the light at 50° to 60° F (10° to 16° C) for eight to ten weeks, then force them into flower.

You can also use amaryllis as cut flowers. Cut them when buds are fully colored but not open. To prevent splitting and outrolling of cut stems, you can hold flowers in 0.125M sucrose (60 oz. sucrose per gal. of water) for twenty-four hours at 72° F (22° C) before shipping.

Amaryllis characteristics and basic program[a]

Cultivar	Color	Average days from pot — Market stage[b]	First flower opening	Average height at first flower — Inches	Centimeters	Average # of flowers — First stalk	Second stalk
South African–grown							
Barotse	Cherry red	29	40	22	56	4.0	4.0
Basuto	Dark red	28	40	22	56	3.8	3.5
Blushing Bride	Rose	28	42	18	46	4.1	3.7
Bold Leader	Red	29	40	18	46	4.2	4.3
Candy Floss	Pink	31	44	22	56	4.1	4.4
Carnival	Red, white	25	36	22	56	4.2	3.6
Cocktail	Red, white	24	38	20	51	4.0	4.4
Desert Dawn	Peach, salmon	32	47	20	51	3.7	3.2
Double Six[c]	Red	25	36	20	51	6.0	6.0
Intokazi	White	24	38	22	56	4.0	3.8
Milady	Pink	26	38	18	46	3.7	3.8
Miracle[c]	Dark red	30	44	20	51	4.5	4.6
Razzle Dazzle	Red, white	25	36	25	64	4.1	4.2
Safari	Light red	25	36	20	51	3.5	4.0
Springtime[c]	Pink, white	35	45	18	46	4.1	4.0
Summertime	Rose, white	28	41	20	51	3.7	3.6
Sun Dance	Bright red	24	35	18	46	3.9	3.8
Wedding Dance	White	24	38	20	51	3.8	3.6
Zanzibar	Bright red	29	42	16	41	4.1	4.1
Netherlands–grown							
Apple Blossom	White, pink	39	55	20	51	4.8	4.3
Clown	White, red	45	60	24	61	3.0	3.7
Flower Record	Rose	33	52	25	64	4.4	4.7
Happy Memory	White, red	48	65	23	58	3.0	3.6
Hercules	Lilac pink	48	62	22	56	4.1	3.9
Ludwig Dazzler	White	41	60	24	61	3.8	3.9
Minerva	White, red	41	57	22	56	4.0	4.4
Orange Souvereign	Orange-red	34	51	22	56	3.9	4.3
Pamelac	Red-orange	26	54	20	51	4.6	4.7
Piquant	Red, white	36	49	18	46	3.9	4.0
Red Lion	Dark red	49	64	22	56	3.2	3.5
Rilona	Salmon-orange	40	61	23	58	3.9	4.2
Scarlet Baby[c]	Red	24	42	22	56	4.0	3.9
Telstar	Red	48	60	18	46	2.8	3.0
Vera	Light pink	34	48	18	46	4.3	4.7
Israeli–grown							
Oscar	Red	52	72	25	64	4.8	4.0
Red Lion	Red	43	67	23	58	3.7	3.6

Source: The information provided is based on trials conducted for one to three years at North Carolina State University.

[a] This is a selected list of cultivars, but many others are available.

[b] Marketing stage for potted amaryllis is when the first floral stalk is 12 in. long. This stage, as well as the flowering stage, usually takes a few days longer than average to reach with early planting; it's reached quicker with later planting.

[c] Cultivar often produces a third flower stalk.

A. A. De Hertogh is a professor at North Carolina State University, Raleigh. July 1997.

Anthurium

Anthurium Production Tips

Gary R. Hennen

After years as a staple cut flower and recent popularity as a pot plant, anthurium has become a popular addition to product lines. Vegetatively propagated, it's usually purchased in tissue culture Stage III or in liners. It's relatively easy to grow, has attractive foliage, and produces long-lasting flowers year-round. Anthurium can be divided into four basic groups:

- *Anthurium andraeanum* cultivars;
- interspecific hybrids between *A. andraeanum* cultivars and dwarf species;
- *scherzeranum* hybrids; and
- foliage anthurium.

The interspecific hybrids are sometimes called Lady Jane types, referring to them as the first widely available cultivar, or as andreacola types.

A. andraeanum, a generally large, somewhat open-structured plant with large flowers, is commonly grown for cut flower production and is sometimes adaptable to pot culture. New cultivars specifically selected for pot culture are more compact. *A. andraeanum* primary flower colors are white, pink, red, red-orange, and green.

In contrast, andreacola cultivars are small to intermediate in size, fuller and more compact. They generally produce smaller but more numerous flowers than andraeanum cultivars. Andreacola cultivars tend to have thicker, dark green leaves and often resist disease. Primary flower colors are white, pink, red, and lavender.

A. scherzeranum, the first widely-cultivated anthurium pot plant, is small and compact. Primary flower colors are white (sometimes with polka dots), pink, and red.

Foliage anthuriums come in many shapes and sizes and represent a minor portion of the total anthurium pot market. However, it should be noted that most foliage anthuriums are durable plants that offer the consumer distinct forms.

Culture

Anthurium prefers evenly moist media, especially when actively growing. Overall, it's better to slightly underwater than overwater. Drying out may cause tip burn, root damage, and reduced growth rates, while overwatering can also cause root damage and sudden yellowing of older leaves.

Anthurium won't tolerate saturated, poorly drained media. Best results are achieved with a 1:1:1 ratio of Canadian peat, composted pine bark (avoid small particle size and too much dust), and perlite or airlite. Avoid vermiculite, except in 4-in. containers. In long-term crops—that is, 6-in. and larger—vermiculite compacts and will waterlog. Soil pH should be maintained between 5.5 and 6.5.

Crop Time

Most pot anthurium are sold in 6- and 8-in. containers, with a smaller percentage in 4- and 10-in. pots. Crop finish times vary depending on cultivar, pot size, and cultural environment. Except in the case of *A. scherzeranum,* consider anthurium a long-term floral crop. Under the subtropical climate of Florida, most 6-in. container crops are finished in eight to ten months using young plants in seventy-two or ninety-eight-cell trays. Scherzeranum is usually grown in 3½- to 6-in. containers and will finish in four to seven months.

Nutrition

Moderate but consistent levels of a complete fertilizer are important. Magnesium requirements in anthurium plant tissue are higher than most foliage crops, especially in warmer climates. Because of the long crop time, pay attention to ensure continued

magnesium availability. Per cubic yard of media, incorporate 10 pounds of dolomite and 3½ lbs. of Hi-Cal lime to balance the calcium and magnesium ratio.

Regular foliar applications of magnesium sources (such as Epsom salts or magnesium nitrate) will help prevent magnesium deficiencies. After twenty-four to twenty-six weeks, a top dressing of dolomite (3 tb. per 10-in. pot) or another magnesium source will help ensure continued magnesium availability. Top dressings of Epsom salts are beneficial but short-lived.

Generally, a 1:1:1 ratio fertilizer is recommended. Avoid high nutrient levels, especially after planting young plants. Liquid fertilizer on a constant-feed program shouldn't exceed 250 ppm nitrogen. On actively growing mature plants, occasional rates as high as 400 ppm nitrogen are acceptable, but such feedings must be alternated with clear irrigations. Tests have shown that plants given frequent doses of 300 to 400 ppm nitrogen actually grow slower, have lighter flower colors, and produce thick, deformed leaves.

Temperature

Anthurium grows best with day temperatures of 78° to 90° F (26° to 32° C) and night temperatures of 70° to 75° F (21° to 24° C). Temperatures above 90° F (32° C) may cause foliar burning, faded flower color, and reduced flower life. Night temperatures from 40° to 50° F (4° to 10° C) can result in slow growth and yellowing of lower leaves. Scherzeranum cultivars require lower temperatures of 68° to 80° F (20° to 27° C) at daytime and 60° to 70° F (16° to 21° C) at night. Anthurium won't tolerate frost or freezing conditions.

Light

Generally, most anthurium grow well at light intensities ranging from 1,500 to 2,500 f.c. Light intensities higher than 2,500 f.c. can improve branching habits (i.e., fullness) but can result in faded flower and leaf color. Some growers use light intensities between 3,600 and 5,000 f.c. during the early crop stages to improve branching, then move the crop to lower light intensities for finishing. Scherzeranum cultivars are best grown at light intensities between 1,000 and 1,500 f.c.

Pest Control

Preventive maintenance programs for mites, snails, slugs, worms, thrips, and whiteflies are important. Whiteflies are especially attracted to new growth and are difficult to eradicate once established. A number of chemicals are effective for pest management; however, cultural conditions and cultivars will determine what you can safely use. Many growers have experienced phytotoxicity on numerous anthurium cultivars from certain pesticides. Never apply pesticides while plants are under any form of stress.

Anthurium andraeanum cultivars are generally susceptible to bacterial blight, *Xanthomonas campestris* pv. *dieffenbachiae*. This aggressive disease starts as foliar necrosis, eventually leading to a systemic infection. Xanthomonas is spread by excessive water, splashing, and contact with infected plants or tools. Susceptible cultivars may perform best under hard cover with drip irrigation and well-drained media. There are no effective chemical controls to prevent blight.

Andreacola cultivars are generally resistant to xanthomonas; however, they're somewhat susceptible to the waterborne fungi phytophthora, rhizoctonia, and pythium. Although many fungicides are effective for these diseases, the best approach is prevention with cultural practices. Keep plants off the ground, provide good ventilation, and avoid overhead irrigation during late afternoon or evening hours.

Gary R. Hennen is president, Oglesby Plant Laboratories, Altha, Florida. December 1996.

Aster

※

Aster Answers: Pot Crop Success

Bob Humm

Asters can be produced in greenhouses year-round using photoperiod control. You can also easily grow them as natural-season fall-pinched or fast crops. They're perfect for overwintering in standard perennial programs.

Media and Fertilizer

Use any well-drained root media; add peat moss to help retain moisture. During production, use a complete N-P-K fertilizer with the majority of nitrogen in the nitrate form and containing extra micronutrients, as in the Peat-Lite Specials. Well-watered and well-fed plants grow best, just like garden mums.

Mum fertilization programs are satisfactory for asters (200 ppm nitrogen in soil mixes; 300 ppm nitrogen in soilless mixes). However, asters are more sensitive to excess soluble salts: Leach with clear water or decrease fertilization levels to avoid high soluble salts. Use only clear water once the flowers begin to open.

Temperature

Soil temperatures of 65° to 68° F (18° to 20° C) are beneficial to root development in the starting area. During growing on, night temperatures should run 62° to 65° F (17° to 18° C), with day temperatures 5° to 10° F warmer. Lowering night temperatures to 58° to 60° F (14° to 16° C) will help intensify colors during the last two weeks of production. In the greenhouse, reduce humidity by heating and venting moist air *before* lowering temperatures to prevent powdery mildew. Flowers on outdoor aster crops will tolerate a light frost.

Spacing

Asters can be spaced pot-to-pot until pinched. However, it's important to place asters at final spacing soon after the last pinch. Otherwise, lower leaves may yellow and turn brown from lack of light or from foliage diseases that occur when plants are crowded. Even before final spacing, leaves shouldn't touch or overlap.

Height Control

Height control on asters is primarily accomplished by proper pinching. Medium or tall cultivars may need B-Nine for added height control. One to two applications of 3,750 to 5,000 ppm B-Nine should be sufficient. Administer the first application after the last pinch when new shoots are about one inch long. Apply an additional application ten to fourteen days after the first, if needed. Don't use B-Nine after buds show color, to avoid clubby flower spray formations. Bonzi sprays of 5 to 10 ppm can also be effective in height control. As with B-Nine, apply at similar growth stages and direct sprays toward plant stems rather than leaves.

Pinching

Pinches create fuller pots. In a greenhouse pot aster program, multiple cuttings with two pinches are generally used. In a natural-season garden aster program, usually only one cutting is used, and multiple pinches are given. In general, the fewer pinches planned, the more cuttings needed to create a full pot.

Give the first pinch when roots are well developed and reach the sides and bottom of the finishing container. This generally occurs ten to fourteen days after planting a rooted cutting. Pinch off enough tip so that only four to six leaves remain. Depending on the original cutting height, this may be either a soft pinch (less than 1 in.) or a hard pinch (more than 1 in.). Rooted cuttings and unrooted cuttings rooted in cell packs for later transplanting naturally become taller during propagation, compared to using unrooted cuttings in direct-stick programs. The first pinch on rooted cuttings will be a much harder pinch than the pinch used for direct-stick unrooted cuttings.

Three to four leaves should remain after subsequent pinchings. Remove a minimum of half inch of new growth. Pinches can be done by hand or with shears and are generally done every two to three weeks.

Take care to pinch all the shoots, or branch height and flowering will be uneven. Avoid deep pinching. Severe pinching or cutting back can lead to vegetative shoots arising from the base of the plant. These shoots often become taller and flower much later than the upper branches, resulting in an unevenly flowering pot. Follow the above guidelines on leaf numbers remaining after the pinch to avoid this problem.

The last pinch date for natural-season crops should be between July 25 (northern latitudes) and August 10 (southern latitudes) to avoid any flowering delay. Later pinches may be used to delay flowering if desired, but plants must be of adequate size, as little regrowth will occur after late pinches. Use late pinches on a trial basis only. Don't pinch after late August.

Florel has been trialed on asters, and it appears to be effective as a replacement for pinching if used early in production. Use Florel only on a trial basis at rates and methods recommended for garden mums.

Table 1. Crop planning for early, shaded crops

Activity	Timing guidelines	Example
Plant rooted cutting	Upon receipt	May 1
First pinch	When ready, about 10 to 14 days after planting	May 15
Second pinch	When ready, about 10 to 14 days after first pinch	May 30
Short days	Two weeks after second pinch	June 14
Flower	Five to six weeks after short days	July 17 to July 24

Note: For 6-in. or gal. pots, use three plants per pot. Plant May 1 to June 1.

Flowering Response

Asters require long days (night lighting 10 P.M. to 2 A.M.) for vegetative growth and short days (blackout for ten to twelve hours, no longer) for flower bud initiation and development. Most varieties will be salable five to six weeks after short days begin.

For greenhouse-forced pots or early-shaded crops, year-round night lighting allows for maximum vegetative growth. Artificial short days (blackout) should be used from March 15 to August 15.

Under natural-season conditions, asters seem to flower best during hot, bright weather. In a normal summer, when hot July and August temperatures tend to delay garden mums, natural-season garden asters generally flower one to two weeks earlier than garden mums. However, in cool and cloudy summers, when garden mums tend to flower earlier, asters are generally delayed and could flower seven to ten days later than garden mums.

Problems

Insects

Whiteflies and thrips are the key insect pests of asters. Regular mum spray chemicals (Avid, Dursban, M-Pede, Marathon, Mavrik, Talstar, Tame, Thiodan) have been effective for such pests.

Diseases

Major diseases found in greenhouse asters are powdery mildew and botrytis. The primary diseases found in outdoor culture are powdery mildew, rust, botrytis, and rhizoctonia foliage blight.

Powdery mildew is characterized by a white-gray, powdery fungal coating on leaf surfaces and stems. It can stunt growth but will rarely kill plants. Rust results in yellow spots on leaf surfaces, with either bright orange fungal growth or raised, brown scalelike pustules under the leaf. Rust can advance quickly, as it's spread by wind and splashing water. It can also kill untreated plants.

Botrytis can cause yellowed and brown lower leaves on overcrowded plants or in areas of high humidity and poor air circulation. Usually, only the lower leaves are affected. Rhizoctonia foliage blight is characterized by brown-black necrotic leaf and stem spots, with leaves dying back from the tips in advanced stages. This disease can quickly kill plants if untreated. However, these aster diseases generally do not spread to an adjacent garden mum crop.

All of the diseases discussed are favored by high humidity and splashing water. Use the following cultural practices to help prevent favorable disease conditions: Use proper spacing, done on time, and avoid overhead watering late in the day. In greenhouses, heat and ventilate to remove excess humidity before lowering night temperatures. This is especially critical when cool nights follow warm, humid days. Remove and destroy any isolated plants that become severely infected with disease so they won't spread disease to other plants. Practice sanitation. Remove debris and weeds from the growing area, as they may harbor diseases. Inspect crops regularly to detect diseases early.

Table 2. Crop planning for natural-season fall crops

Crop	Container size	Plants per pot	Plant date	Number of pinches	Approximate spacing (in.)
Normal	8 by 5 in.	1	Late May	3	20 by 20
Normal	1 to 1½ gal.	1	Early June	2	18 by 18
Normal	8 by 5 in.	1	Mid-June	2	18 by 18
Normal	1 to 1½ gal.	1	Late June	1	16 by 16
Normal	8 by 5 in.	1	Early July	1	16 by 16
Normal	8 by 5 in.	2	Mid-July	1	16 by 16
6-in. fast crop	6 to 6½ in.	1	Mid-July, late July	None	12 by 12
4-in. fast crop	4 to 4½ in.	1	Late July, early August	None	8 by 8

Note: Plant dates are based on starting with rooted cuttings and are based on Midwestern/Eastern region growing conditions.

Chemical programs are most effective when used along with proper cultural practices; they can't effectively perform alone. Generally, if good cultural practices are followed, greenhouse pot asters need little chemical treatment. Spray as needed for botrytis or powdery mildew control.

With outdoor aster production, a preventative chemical spray program is critical. Using approximate dates of July 15, August 15, and September 15, apply Bayleton (Strike) to suppress powdery mildew and rust. (Don't apply more frequently than thirty-day intervals.) Use Chipco 26019 (or Cleary's 3336) sprays between Bayleton sprays to suppress botrytis, rhizoctonia, and other leaf-spotting fungi. To avoid potential plant injury, don't mix Bayleton with other fungicides. If diseases persist, try alternative chemicals, and review your cultural practices.

Table 3. Crop planning for perennial crops

Activity	Timing	Cover	Uncover
Plant rooted cuttings	August 15 to October 15	Mid-December	March 1

Note: For overwintering in 1801 packs, quarts, or gallons.

Bob Humm is seasonal and broker products manager, Yoder Brothers Inc., Barberton, Ohio. April 1997.

Azalea

Proper Cooling and Forcing Produce High-quality Azaleas

Terril A. Nell and Ria T. Leonard

Poor azalea flowering is usually blamed on branching problems or improper environmental conditions, but University of Florida research revealed that failure of buds to open is usually caused by the wrong cooling and forcing conditions. A high-quality 6-in. azalea plant should have forty or more buds, with eight open at the time of sale and the remainder showing color.

Precision cooling allows for accuracy in cooling and forcing programs. In these programs, growers move azaleas from greenhouse to cooler when buds are well-developed. Azaleas remain in the cooler long enough to meet the cold requirements of the bud (usually six weeks), just as if plants remained under natural conditions during the winter. Cooler light and temperature conditions are so different from those used in an overwintering greenhouse that special care is required to ensure best results.

Lighted Cooler

Lighted coolers produce plants with uniform flower bud development and no leaf necrosis or browning. Maintain cooler temperatures for six weeks at 42° to 48° F (6° to 9° C), depending on time of year and variety. Cool varieties prone to leaf necrosis (Dorothy Gish, Gloria and White Gish) at 42° F (6° C). Varieties that rarely have leaf problems (Solitaire and Prize) can handle 48° F (9° C). In most cases, temperatures can be reduced to 35° to 38° F (2° to 3° C) to eliminate any risk of leaf problems and to foster more uniform flowering.

Time from removal from the cooler to flowering will be delayed slightly as cooler temperature is decreased. Plants fertilized heavily during the final three to six weeks of production (before going into the cooler) are more likely to exhibit leaf problems than plants grown with reduced fertilizer applications.

Maintain light level at 30 f.c. for twelve hours daily. The source of light, fluorescent or incandescent, doesn't appear to affect a properly cooled azalea. You can save energy by mounting the ballast for fluorescent lights outside the cooler, thus removing the majority of heat from fluorescent lamps. Mounting the fluorescent light ballast externally makes fluorescent fixtures more economical by making them more efficient and reducing heat. Extending the lighted period beyond twelve hours hasn't improved bud uniformity or leaf quality. Maintain humidity between 80 and 90% to avoid rapid drying of plants and desiccation of leaves. In some coolers, you may need humidifiers to achieve desirable humidity levels.

Care During Cooling

Depending on cooler humidity and air movement, plants will require one or more waterings during the cooling period. Water plants thoroughly before placing them in the cooler, and monitor soil moisture approximately twice weekly. Determining the proper time to water in the cooler is very difficult. Cold soil feels wet even if the media is beginning to dry out. It's best to select several pots and judge the time to water based on the container weight. Also, experience and good records are critical. Maintain records to show when you water each group or variety of plants.

Consider air movement and air quality in operating azalea coolers. An azalea cooler should provide approximately two air exchanges per hour. The cooler should be designed so the air exchanger doesn't discharge directly onto the plants. The high airflow directly in front of the air handling unit causes leaf desiccation and browning, often rendering plants unmarketable. The presence of ethylene can cause green leaf drop in azaleas, so don't store fruits and vegetables in the same cooler with azaleas, and keep the cooler free of dead and decaying leaves and flowers.

Dark Coolers

The conditions for the dark cooler (humidity, air movement, air quality, and watering practices) are the same as for the lighted cooler. In most cases, comparable quality plants can be produced with either cooler, provided you carefully monitor temperature and maintain it at proper levels. The dark cooler doesn't provide protection against leaf problems on sensitive varieties. Maintain the dark cooler at 35° to 38° F (2° to 3° C), or leaf problems may become prevalent. The greatest problem observed with the dark cooler is failure to maintain sufficiently low temperatures. In most of these problem situations, the grower thought the cooler was being maintained at proper temperatures, only to find that the actual temperature was several degrees warmer. Allowing the temperature to rise 3° to 4° F in a dark cooler can result in severe leaf damage in a short time period. Using a thermograph provides a good overview of the temperature in the cooler.

Gibberellic Acid

Applying gibberellic acid will provide the dormancy breaking treatment just as effectively as a cooler treatment if the application rate, application timing, and bud stage are correct. With the use of gibberellic acid, you can spray the plants in greenhouses, eliminating costs of cooler operation and labor to move plants in and out of the cooler. Gibberellic acid from GibGro4LS, Agtrol Chemical Products, Houston, Texas, had only been labeled for use on azaleas only in Florida in 1996, but the manufacturer added gibberellic acid to the national label in early 1997.

Proper gibberellic application requires that buds are more developed (Stage 6, style elongated and closed) than plants receiving cold treatments. Once the plants have reached Stage 6, weekly applications of gibberellic acid at 500 ppm for six weeks, or until flowers are open, are necessary to overcome dormancy. In some situations, growers have successfully cooled plants for four weeks, then made three applications of gibberellic acid rather than cooling plants for two additional weeks.

Forcing Conditions

Once you remove azaleas from a cooler, you must maintain certain conditions to optimize bud development and opening. In gibberellic acid–treated plants, the dormancy-breaking treatment and forcing are combined. Best forcing temperatures are 65° to 68° F (18° to 20° C) during the night and maximum day temperatures of 75°to 80° F (24° to 27° C). Plants will flower faster at warmer temperatures, but flower color is poor, and longevity is reduced compared to cooler temperatures. Light levels of 2,500 to 4,000 f.c. allow for good bud development. Many small buds don't develop at lower light levels, so flowering isn't uniform throughout the plant canopy. Proper watering is critical in forcing azaleas. Overwatering leads to root damage and disease, while drying delays flowering and often results in bud abortion. No fertilization applications are necessary during this forcing phase. Plants watered improperly during forcing don't last in the interiorscape.

Terril A. Nell is professor of floriculture and Ria T. Leonard is research assistant, University of Florida, Gainesville. July 1996.

Baskets

❈

Great Baskets: On a Budget or Upscale

Kerstin Poehlmann

If you want your consumers to be impressed when they walk into your nursery, nothing is more effective for grabbing their attention than mixed baskets. Mixed baskets have gained great popularity among consumers over the past few years, and the market has responded accordingly, becoming much more competitive. A wide assortment of mixed baskets isn't a specialty anymore—it's a necessity. So, how do you stand out from the crowd and still remain competitive? How do you design, grow, and display mind-blowing baskets and get your fair share of the pie? Read on, and you'll see that great baskets don't have to be complicated, hard to grow, or expensive. No matter what your budget is, get ready to get your hands dirty—it will be well worth it.

Growing Tips

The first decision you have to make when planning a basket is whether you want to plant your starter material right into the basket or grow it in 4-in. pots first and transplant it into baskets later. There are pros and cons for both.

A basket grown from plugs or liners looks more natural because plants grow together for their whole lives and blend in better. It also reduces labor costs, because it saves the extra step of transplanting 4-in. pots into baskets.

On the other hand, growing individual varieties in 4-in. pots before transplanting into baskets allows you to try different designs and actually see the end result before transplanting. It's harder to imagine what the basket will look like if you start from liners. Another advantage of growing in 4-in. pots is that they require less space in the beginning and are easier to handle if they need to be moved. It also eliminates the risk of having a gap in the basket if one of the starter plants dies, and it lets you customize designs and use leftover plants in your combinations. Pest control becomes easier, as you can treat only the crops with problems.

Certainly, growing in 4-in. pots and transplanting later would seem to be the best solution, but the advantages of a nicer, more natural appeal and reduced labor costs aren't to be underestimated. You'll have to decide what your priorities are.

Soil

Soil requirements for mixed baskets are the same as for single baskets. Use a peat/perlite mix with good drainage. This is especially important to avoid waterlogged soil and root rot if you grow in plastic baskets. If you use moss or coco fiber baskets and use plants that naturally need a lot of water, you should consider adding a water-retaining agent to your soil mix, especially if you live in a hot climate with strong winds.

Fertilizer

During production, fertilize with a well-balanced fertilizer. A 20:5:15 formulation with average micronutrients and slightly high iron works well. Iron is a must if growing trailing petunias, but it's also recommended for most bedding plants. Apply 200 to 250 ppm nitrogen with constant feed or 250 to 300 ppm nitrogen with periodic feed for most varieties. If you're producing trailing petunias, increase to 300 to 350 ppm constant feed or 350 to 400 ppm periodic-feed nitrogen. If plants show any signs of chlorosis, supplement with chelated iron. Chelated iron can also be used as a foliar feed to green up the foliage quickly. Look at it as an emergency first-aid procedure because it's quite expensive, but it certainly works.

To keep consumers from bringing back baskets with pure yellow foliage, always add slow-release fertilizer to your soil mix before planting baskets. Use it more generously if planting trailing petunias. If you have a retail operation, offer fertilizer for sale right next to your baskets, and remind your customers to feed their plants well.

Pest control

Inspect your baskets regularly and thoroughly for pests. Because you're growing a variety of plants together in one basket, you're offering "something for everybody," as different plants attract different pests. Spray preventively if you've experienced many pest problems in the past. Otherwise, just watch closely as you would with individual crops. When pest control is necessary, make sure that none of the varieties in the basket are sensitive to the particular pesticide. Once plants in the combo get large, pests are harder to treat because they can hide in the thick foliage, so make sure you start looking for them early.

To prevent soil-borne fungal diseases, drench with a broad-spectrum fungicide immediately after planting, whether you plant liners or transplant 4-in. pots. Make sure you combine only plants with similar water requirements. Otherwise, some varieties will be stressed and become more susceptible to pythium.

Other cultural requirements aren't different from producing individual varieties. Make sure the plants in your combos have similar light and temperature requirements and similar growth vigor. Otherwise, nature's law will take over: survival of the fittest. Needless to say, Supertunias would outperform alyssum. Likewise, in a geranium–African violet combo, one of the two wouldn't do well, depending on which plant prefers the light conditions available.

Design Tips

Designing great baskets is easier than it seems. Following these guidelines will ensure success.

Color

The easiest way to make an impact with mixed baskets is by effectively using color. Color catches the eye first. Decide whether to create a color contrast or use a tone-in-tone combination. Color-contrast combinations use complementary colors such as yellow/orange tones mixed with blue/purple colors. If you don't know which colors are complementary, buy a color wheel for a few dollars at an artist's store. Complementary colors are displayed opposite each other. Tone-in-tone combos use similar shades and tones of one color or colors that are close to each other on the color wheel. The easiest high impact color combination is the combination of the three primary colors: red, yellow, and blue.

Texture

The best way to add texture to a combo is to use component plants such as plectranthus, mentha, glechoma, ivy, vinca, or different colors of *Helichrysum petiolare* (licorice plant). Their foliage colors and textures give baskets the look of a "living bouquet." To accentuate with component plants without over-powering the flowering plants, a ratio of three to four flowering plants to one component plant is a good rule of thumb.

Proportion

The right proportion contributes to the overall appeal of the basket. The basic guideline is for the plant to be twice as high as the visible part of the basket. Using plants with different heights also helps achieve balanced proportions. Plant upright varieties in the middle; use mounding plants next; then put trailing ones on the edges.

LIBRARY PERSHORE COLLEGE AVONBANK PERSHORE WORCESTERSHIRE WR10 3JP

Containers

The container influences both the design as well as the budget. The choices are endless and reach from basic plastic baskets to coco fiber and moss baskets to specialty items such as terra-cotta baskets, wooden baskets, wall baskets, and other uniquely shaped items.

If budget is a main concern, the standard plastic basket is a good choice. Draw the focus away from the basket by using trailing plants such as petunia, Supertunia, Surfinia, and ivy geraniums. They'll soon cover the basket, and the only visible part of the plastic container will be the hanger. Choose a green basket because it will blend with the foliage and leave the focus on the plants.

If you want to use a more natural product but still be budget-conscious, choose coco fiber baskets instead of moss baskets, which are usually slightly more expensive. Moss baskets seem to be customers' choices if the material is visible, but again the fiber material can be covered with trailing plants. Plant performance is very good in both materials.

Moss baskets are a great choice for medium-priced to upscale baskets. They're very attractive by themselves and allow you to sell baskets early when plants are still smaller and more of the basket is visible. They can easily be planted on the sides, creating a lush, generous mixed basket.

If you want to create mind-blowing upscale combos, don't try to save money on the container. Use what you like best. Elegant terra-cotta or rustic wooden baskets create an upscale appearance by themselves and add value to any plant combo.

Kerstin Poehlmann is director of sales and marketing, EuroAmerican Propagators, Bonsall, California. January 1998.

Foolproof Recipes for Hanging Basket Success

Kerstin Poehlmann

The following suggestions have something for everybody. Whether budget-conscious or upscale, these ideas will make your customers come back for more because they include budget, design and plant requirements for long-lasting, great baskets.

The budget-conscious

- Combine one pink impatiens, one verbena Tapien Pink and one soft purple brachycome Ultra in a 10- or 12-in. plastic basket—a great tone-on-tone combo.
- Mix one scaevola New Wonder, one geranium Lilac Cascade, one pink verbena, and one variegated helichrysum Licorice in a 12-in. plastic basket—a special look in harmonizing colors.

- Put three verbena Tapien Blue-Violet and one helichrysum White Licorice in a 12-in. plastic basket—a striking color contrast.
- Use one red zonal geranium, one helichrysum Golden Beauty, and one angallis Skylover in a 12-in. coco fiber basket—a vivid combination of the three primary colors.

Nice, but not pricey

- Try one scaevola New Wonder, one geranium Lilac Cascade, one petunia Supertunia Purple Sunspot, and one pink impatiens with vinca vine in a 14-in. moss basket—harmonizing colors nicely presented.
- Mix petunia Surfinia Pink Vein, petunia Surfinia Blue Vein, carnation Cinnamon Red Hots, verbena Tapien Pink, and brachycome Ultra in a 14-in. moss or coco fiber basket—pastel colors in a lush basket.
- Plant one orange gazania, one heliotrope, one lobelia Compact Royal Jewels, one helichrysum Golden Beauty, and one plectranthus in a 14-in. moss basket—unusual color contrast combo.

The money makers

It's not just the choice of plants that makes your baskets look like a million dollars; it's also the container and the presentation. Try these ideas for upgrading:

- Plant any of the above combinations in a terra-cotta or wooden basket, and you'll get more than your money's worth.
- Add an elegant grass and a lesser known plant like a rusellia to give your combo an elegant, exotic look.
- Don't underestimate the power of presentation. Arrange your upscale baskets in a designated area decorated with appropriate props and accessories such as ceramics and statues. Create seasonal displays, for example, using pumpkins, scarecrows, and straw for combos with fall crops.
- Take advantage of special occasions throughout the year to get extra sales. Create special combos in red, white, and blue for the Fourth of July, and decorate them with flags. Use white Marguerite daisy or zonal geranium with verbena Tapien Blue-Violet, and verbena Temari Bright Red. Plant combos of red and white zonal geraniums with bacopa Snowstorm and verbena Temari Bright Red for Mother's Day, and decorate them with bows.
- Combine the new, frost-hardy, autumn-colored component plants with traditional fall crops such as mums and pansies. This new concept adds value to fall flowering plants and extends your sales season. Present them with fall props for an extra upgrade in value.

Kerstin Poehlmann, EuroAmerican Propagators, Bonsall, California. January 1998.

Herb Baskets

Teresa Aimone

Why not use herbs for 10-in. (or even larger) baskets? The different textures, fragrances, and plant heights make them perfect for arranging a miniature herb garden. Make up baskets for shade-loving and sun-loving areas, and include information on care and use. There are herbs used for their edible flowers, for tea, potpourri, dyes, particular color schemes (such as gray or silver gardens), and, of course, for cooking.

Plant three to five large plugs per basket. Keep plants well-watered and fertilized. Some examples of plant combinations include: chives (center for height) with parsley, sage, and thyme; chamomile with lemon balm, mint, and lavender; and dill with sweet basil, French tarragon, and oregano. Experiment with the wide variety of herbs available.

Teresa Aimone was a regional specialist Southeast, S&G Seeds, Coppell, Texas. April 1998.

Begonia

Beautiful Begonias

Tom Linwick

Begonia semperflorens has become an important crop for most plug growers. Because begonias take more time to produce than other bedding plants, many growers would rather buy in plugs than sow them. Begonias have been steadily increasing in popularity as landscapers and homeowners find that they have good garden performance and are low-maintenance bedding plants.

With all of the recent breeding work in begonias, the production time has been reduced to between eleven and twelve weeks depending on the time of year. Seed quality is usually excellent, so germination rates are high. Most plug growers like to grow begonias because they are reliable as far as timing, are relatively easy to produce, and can be held a bit longer than other types of plants.

Even with improvements in seed quality, begonias still require the proper cultural conditions to produce a quality crop. There are slightly different temperature requirements for germinating pelleted seed versus raw seed. In the following information, I will outline some of the steps for successful begonia production.

It is best to use a seeding mix that is low in nutrients, with an EC value of about 1.2. The pH should be between 5.5 and 6. Before sowing, it is recommended to water seed flats and plug trays with a fungicide to eliminate soil-borne fungus. Both raw seed and pelleted seed need light to germinate, so do not cover seeds after sowing. The recommended temperature for germination is 72° F (22° C) for raw seed and 79° F (26° C) for pellets. The higher temperature helps to break the coating on the pelleted seed. Humidity is very important in the initial germination, so try to keep it as close to 100% as possible. Many growers who germinate on the bench will cover the seed trays with plastic to keep the humidity levels up for the first five to six days after sowing. Those who use a germination chamber will leave them in the chamber for approximately six days or until the seed is starting to germinate. Then the seed flats or plug trays are moved into the greenhouse and grown at 68° F (20° C).

About two weeks from sowing, all seeds will have germinated and will be visible to the naked eye. At this point the seed flats or plug trays can be run a little drier to help get the plants rooted in and prevent a buildup of algae on the seed tray surface. When watering young begonia seedlings, do not water with too much force since newly germinated seedlings can be disrupted by the water, resulting in roots being torn away from the soil. Once plants are well germinated, or at about fourteen days, feed with a complete fertilizer at about 75 to 100 ppm. When the plants are three to four weeks old, they require moderate feedings with a complete fertilizer at 150 ppm.

At five to seven weeks old, depending on the temperature at which they are grown, plants should be ready to be transplanted. We recommend keeping the temperature up to 66° to 68° F (19° to 20° C) for the first two to three weeks after transplanting. Once plants are well established, the temperature can be dropped to about 64° F (18° C), and when the plants need to be hardened off, the temperature can be lowered to 61° F (16° C). Supplemental lighting is beneficial if light levels become too low. In the early sowings during the winter months, you should supply 50W/m before transplanting. The use of growth regulators is not required, but Cycocel can be used at 300 to 500 ppm. If you use Cycocel, do not use it until the plants are established in their final containers since it may keep plants from filling out plug flats and even the finished container.

Begonias will continue to be an important crop for plug and bedding plant growers. Due to their excellent garden performance and low maintenance, begonias are very attractive to both landscapers and homeowners. Since plug growers have become very good at begonia production, finish growers don't need to propagate them anymore.

Tom Linwick is technical representative, Daehnfeldt Inc., Duvall, Washington. December 1997.

Bulbs

Pot bulbs

A. A. De Hertogh

Ornamental flowering bulbs as potted plants has increased in popularity and in diversity during the past thirty years. While some are marketed in the sprout stage as growing plants or as bedding plants for transplanting into the landscape, others are sold in the bud stage as flowering potted plants (Table 1). The wide range of bulb sizes makes them suitable for containers ranging from 3-in.-diameter pots to 8-in. bulb pans (Table 2). The species diversity among ornamental flowering bulbs—geophytes—extends the market season from seasonal with spring-flowering bulbs such as tulips, to year-round with Asiatic and Oriental hybrid lilies.

Flowering bulbs' growth and development requirements are classified in two forcing groups: rooting room and non-rooting room. All rooting room bulbs must be completely rooted and cooled under refrigeration and then placed in the greenhouse, while most non-rooting room bulbs are stored until planted and then rooted under greenhouse conditions.

Production Phase

The production goal for bulb growers is to produce disease-free, true-to-type bulbs that, when properly programmed, will flower without problems. Forcers can ensure success by purchasing the proper bulb size (Table 2) from the best production sources for each market use (Table 1). Since bulb production areas are normally very distant from forcing facilities, it's essential that all flower bulbs be properly packed and transported.

Programming and Greenhouse Phases

Optimum postharvest storage requirements depend on the bulb species being forced. All rooting room bulbs require warm (60° to 70° F/16° to 21° C) temperatures immediately after harvest until the flower(s) reaches a stage for either precooling or planting (non-precooling). They should be packed in well-ventilated trays and transported in containers at 60° to 65° F (16° to 18° C). Depending on the cultivar and bulb type, they require low temperatures for ten to twenty weeks to properly cool them.

Most non-rooting room bulbs, however, have very specific packing and transportation requirements depending on species. For example, Asiatic and Oriental

Table 1. Marketing seasons and potted plant uses for ornamental flower bulbs (geophytes)

Production system	Bulb species	Flowering pot plants	Growing plants	Bedding plants	Marketing season
Rooting room	*Allium karataviense*	X	-	X	Apr-May
	Crocus	X	X	-	Jan-Mar
	Hyacinthus (hyacinth)	X	X	X	Dec-Apr
	Iris danfordiae	X	X	-	Dec-Feb
	Iris reticulata	X	X	-	Jan-Feb
	Leucojum aestivum	X	-	X	Feb-Mar
	Muscari armeniacum	X	X	X	Jan-Mar
	Narcissus (daffodil)	X	X	X	Dec-Apr
	Scilla tubergeniana[1]	X	X	X	Jan-Mar
	Tulipa (tulip)	X	X	X	Dec-May
Non-rooting room	Amaryllis (hippeastrum)	X	X	X	Nov-May
	Anemone coronaria	X	-	X	Feb-May
	Astilbe	X	-	X	Mar-June
	Caladium	X	-	X	Apr-July
	Convallaria (lily-of-the-valley)	X	-	X	Year-round[2]
	Dahlia	X	-	X	Apr-June
	Freesia	X	-	X	Year-round[2]
	Lilium (lily, colored)	X	-	X	Year-round[2]
	Lilium (lily, Easter)	X	-	-	Mar-Apr
	Narcissus (paperwhite)	X	X	-	Nov-Apr
	Ornithogalum dubium	X	-	-	Jan-Mar
	Oxalis species	X	-	X	Nov-May
	Ranunculu	X	-	X	Mar-Apr
	Zantedeschia (calla lily)	X	-	X	Nov-July

[1]Officially classified as *S. mischtschenkoana.*
[2]Highly dependent on prevailing weather conditions.

hybrid lilies must be stored at 35° to 41° F (2° to 5° C) immediately after being harvested for at least six to eight weeks. Others, like amaryllis (hippeastrum), require a minimum of eight to ten weeks postharvest storage at 48° to 55° F (9° to 13° C) before planting. Paperwhite narcissus requires two to three weeks at 63° F (17° C) after initially being stored at 80° to 86° F (27° to 30° C) to form flowers. Specific programming treatments aren't only species specific but also cultivar specific and must be carefully managed to produce a marketable crop.

Planting Media

The planting medium for each bulb type is very critical. All bulbs require a medium that's sterile, well-drained, and low in soluble salts with a pH of 6 to 7. The forcer can prepare the medium or purchase commercial mixes like Ball Grower Mix Number 3, which is excellent for most rooting room bulbs, and Sunshine Mix Number 4, which is excellent for amaryllis, astilbe, and Asiatic and Oriental hybrid lilies.

Watering

Keep the planting medium moist but not overwatered. This requires daily inspections and is dependent on forcing facilities and prevailing weather conditions.

Light

With a few exceptions such as Easter lilies, most flowering bulbs aren't markedly affected by photoperiod. Plant quality is, however, affected by light intensity. For example, convallaria requires low light conditions, but freesia and all lilies prefer high light intensities. Most rooting room bulbs used as potted plants should be forced at medium light intensities.

Fertilization

The only rooting room bulbs that require fertilization during forcing are *Allium karataviense* and tulips. Non-rooting room bulbs that require fertilization include *Anemone coronoria*, astilbe, caladium, dahlia, freesia, Dutch iris, all lilies, oxalis, and zantedeschia. In general, only $Ca(NO_3)_2$ and KNO_3 are needed, but some prefer a complete (NPK) fertilizer.

Plant Growth Regulators

Florel is useful for controlling stem topple and total plant height in potted hyacinths and narcissus (daffodils). Depending on labels, A-Rest, Bonzi, and Sumagic can be used to control height in potted tulips, dahlia, freesia, and all lilies.

Greenhouse Temperatures

Most potted bulb plants should be forced in a 60° to 65° F (16° to 18° C) night/65° to 70° F (18° to 21° C) day temperature greenhouse. *Leucojum aestivum* should be forced cooler (50° to 60° F/10° to 16° C), and amaryllis, caladium, convallaria, and oxalis should be forced warmer (70° to 75° F/21° to 24° C day/night).

Disease and Insects

Diseases affecting flower bulbs during forcing are species specific. For example, lilies can develop a root rot complex, and zantedeschia is susceptible to erwinia soft rot. However, most flower bulbs can be affected by botrytis. Thus, it's important to start with high quality bulbs and a well-drained planting medium to help prevent disease. Proper management of programming and greenhouse environment can reduce diseases.

Aphids and thrips affect many flower bulbs. Use an integrated pest management (IPM) program, including scouting for all pests with the goal of using few pesticides.

Marketing and Consumer Phases

For consumers to get the longest life from potted flowering bulbs, forcers have to market them in the proper development stage. This varies with each bulb species. In some cases, bulbs can be sold directly from the rooting room. Others should be forced in the greenhouse until the first sign of flower color or one to three flowers open.

Table 2. Number of flower bulbs (varying sizes) for a range of pot sizes

		Pot size (diameter)				
Bulb species	Bulb size	3 inch (7.5 cm)	4 inch (10 cm)	5 inch (12.5 cm)	6 inch (15 cm)	8 inch (20 cm)
Allium karataviense	12/up cm	—	—	—	3	6
Amaryllis	30/upcm	—	—	—	—	1
	28-30 cm	—	—	—	1	—
	26-28 cm	—	—	—	1	—
	24-26 cm	—	—	—	1	—
Anemone	3-4 to 6/up cm	—	—	—	4-6	—
Astilbe	Single eye	—	—	—	3	—
(false spirea)	Triple eye	—	—	—	1	—
Caladium	Jumbo	—	—	—	1	—
	No. 1	—	1	—	3	5
	No. 2	—	2	3	4-5	—
Convallaria	No. 1	—	4-5	5-7	8-12	—
(lily-of-the-valley)						
Crocus	10/up cm	3	5-6	7-9	10-12	—
Daffodils (see narcissus)						
Dahlia	Division	—	—	—	1	3
Freesia	5/up cm	—	4-6	5-7	6-10	10-15
Hyacinthus (hyacinth)	18-19 cm	—	1	—	3	5-7
	17-18 cm	—	1	—	3	5-7
	16-17 cm	—	—	3	5	8
	15-16 cm	—	—	3	5	—
Iris reticulata and *Iris danfordiae*	6/up cm	5-6	9-12	14-16	18-20	—

Forcers should provide consumer care tags for marketable potted plants. Proper handling by consumers will maximize plants' lives and increase satisfaction with the product.

For specific details, see the *Holland Bulb Forcer's Guide*, available from Ball Publishing.

A. A. De Hertogh is professor, North Carolina State University, Department of Horticultural Science, Raleigh, North Carolina. November 1995.

Table 2. (Continued)

Bulb species	Bulb size	Pot size (diameter)				
		3 inch (7.5 cm)	4 inch (10 cm)	5 inch (12.5 cm)	6 inch (15 cm)	8 inch (20 cm)
Leucojum aestivum	14-16 cm	—	—	3	3	8
Lilium (lily)	18-20 cm	—	—	—	1	—
	16-18 cm	—	—	—	1	—
	14-16 cm	—	—	—	1	—
	12-14 cm	—	—	1	1 or 3	—
	10-12 cm	—	—	1	3	—
L. longiflorum (Easter lily)	Most sizes	—	—	1	1	—
Muscari	9-10 cm	—	6	8	10	—
Narcissus (daffodil)	DNI	—	—	—	3	5
	DNII	—	—	—	3	5
Narcissus (miniature)	DNIII	—	2	3	5	7
Narcissus (paperwhite)	14/up cm	—	—	3	5	7-9
Ornithogalum dubium	4-5 cm	—	1	—	—	—
Oxalis species	Top-sized	2	3-4	5-6	6-8	—
Ranunculus	5-6 cm	—	1	—	3	—
Scilla tubergeniana (S. mischtschenkoana)	6-7 cm	3	5	—	—	—
Tulipa (tulip)	12/up cm	—	3	5	6-7	9-10
Zantedeschia (calla lily)	2 in. (diameter)	—	1	1	3	3

Caladium

Caladiums

Lynn P. Griffith Jr.

Caladium, which is fairly easy to grow, is produced in small pots, typically three, four, and six inches. It's a quick-turn crop, usually completed within five to eight weeks. Its heart-shaped, multicolored foliage makes the plant popular for Easter and Mother's Day pot sales. The market is year-round, though, with emphasis on spring and summer sales.

Varieties

Today about a hundred or so varieties are in commercial culture, with about twenty varieties making up the main body of the market. Some prominent cultivars include Candidum (white with green veins), Frieda Hemple (predominantly pink), Fannie Munson (predominantly red), and the lance-leaf variety, White Wing. Varieties that tolerate indoor conditions include Lord Derby, White Christmas, Fire Chief, Red Flash, Carolyn Whorton, Poecile Anglais, Sea Gull, Scarlett Beauty, and Aaron.

Commercial cultivars known as the fancy-leaved caladiums are normally hybrids of *Caladium* x *hortulanum*. Some are also *C. picturatum* cultivars, the modern lance- or strap-leaf varieties. These are shorter plants with smaller leaves that are more ruffled along the edges. Most caladium tuber (bulb) production today occurs in the moist, organic soils around Lake Placid in Highlands County, Florida.

Propagation

Nearly all caladiums are produced from tubers, commonly called bulbs, about 95% of which are field grown in Central Florida. They can be grown from seedlings, but plants from seed are slow to finish and generally aren't produced. Propagation has been done from tissue culture from time to time, but cost factors and varietal instability have been limiting factors in tissue-culture caladiums.

The tubers are harvested from the field and stored for eight weeks. Tubers shouldn't be exposed to temperatures below 60° F (16°C). When tubers arrive from the supplier, they should feel firm, rubbery, and somewhat sweaty. If they're spongy, it means they've been exposed to temperatures below 60° F. The tubers should have been stored at 70° F (21° C) for eight weeks and not less than six weeks.

Many growers prefer to gouge or scoop out the dominant eye of the tuber in order to get more shoots of uniform size. This results in a somewhat fuller plant, though with smaller leaves, and gives a slight delay in sprouting.

Culture

Being strictly tropical, caladiums should be grown under warm conditions, ideally 80° to 90° F (27° to 32° C) and a minimum of 70° F. Bottom heat is common in northern climates. Research has shown that increasing the temperature from 70° to 90°F (21° to 32° C) speeds up the crop only a little bit, so in more temperate climates, you can stay at the lower end of the temperature range. Minimum night temperatures should be 65° F (18° C). Plants suffer injury at 55° F (13° C), and most are killed when exposed to 35° F (2° C).

Light levels are usually 60 to 80% shade, equivalent to 2,500 to 5,000 f.c. Color patterns of varieties change with light level, and you should experiment with different light levels to find the ideal range for your cultivars. Candidum is frequently grown at 1,000 f.c., whereas red and pink varieties are produced at the brighter end of the range.

Crop times vary somewhat by planting date and number of tubers per pot, but most plants can be finished in five to eight weeks. The number of tubers varies with variety and cost factors and whether tubers are scooped. A 6-in. pot is frequently planted with three to five #2 tubers. Smaller pots may have only one or two tubers, especially when scooped. Planting depth should be 1 to 1½ in. Caladiums are tolerant of a wide range of potting media. Two parts peat to one part sand is used in some areas, as are mixtures of two parts peat, two parts bark, and one part sand. Shoot for a pH of 5.5 to 6.5.

Nutrition

Unlike many foliage plants, caladiums prefer a 2-2-3 ratio of $N-P_2O_5-K_2O$ or a 1-1-1 ratio. Granular 6-6-6 or slow release 14-14-14 are frequently used at about one teaspoon per 6-in. pot. High nitrogen applications are discouraged, as plants tend to grow very weak and stretched, with poor color. Growers using liquid fertilizer frequently use a constant liquid feed of about 150 ppm nitrogen and 150 to 200 ppm K_2O.

The caladium is one of only a few foliage plants where nutritional factors and deficiency symptoms have been well researched. The plant has relatively high requirements for potassium, magnesium, calcium, and boron. See the following "Disorders" section for nutrient deficiency symptoms.

Diseases

Like most other aroids, caladiums are susceptible to dasheen mosaic virus. Growers have done a good job reducing or eliminating the virus from stock; therefore, dasheen isn't a major problem in caladiums today.

Root knot nematodes are a significant problem for tuber producers, but soil fumigants and hot water dips (122° F/50° C for thirty minutes) are effective against nematodes. *Fusarium solani* causes a common chalky tuber rot. A thirty-minute soak in a thiophanate methyl fungicide is helpful. In hot weather, southern blight, caused by the fungus *Sclerotium rolfsii,* can kill plants. The fungus is characterized by white threads growing on soil and plant surfaces, with white to tan fruiting bodies (sclerotia) that are quite visible. Sprays or drenches with either Terraclor (PCNB) or the insecticide Dursban (chlorpyriphos) are effective. Tubers will also rot when exposed to temperatures below 50° F.

Insect and Mite Pests

Aphids and thrips are the most common insect problems on caladium. Aphids, of course, feed on new leaves and cause sucking injury and puckering of foliage. Thrips frequently create rasping injury in the unfurling leaf, and damage is often worse on one side of the leaf than the other. Malathion is frequently sprayed to control these pests, as are Orthene (acephate) and Mavrik (tau-Fuvalinate). Mealybugs, which can attack tubers or potted plants, are generally controlled with Dycarb (bendiocarb) or Cygon (dimethoate).

Two-spotted mites can be a problem in warm weather, giving you small, white specks on foliage from their feeding injury. Sprays of Pentac (dienochlor), Avid (abamectin), or insecticidal soaps are useful. Silverleaf whiteflies have become a production problem in recent years in many areas. The best control measures vary with the population of whiteflies, but useful materials include the neem extracts (such as Azatin), combinations of Orthene (acephate) and Talstar (bifenthrin), or Marathon (imidacloprid).

Disorders

Many caladium cultivars, especially white ones, show foliar burn symptoms when drought-stressed or exposed to high light. Large, oval lesions form in the interior part of the leaf, starting out with a translucent appearance and usually ending up as a tan lesion or a hole in the leaf.

Nitrogen-deficient caladiums show chlorosis and reduced leaf size. When lacking in phosphorus, plants are small and stunted. Potassium deficiency is characterized by marginal chlorosis and foliar lesions, especially in older leaves. Calcium-deficient plants have reddish brown spots on undersides of leaves near petioles. Caladiums lacking magnesium have chlorosis and holes in the older leaves, while iron-deficient plants have yellow new leaves with green veins. When lacking manganese, plants turn an odd yellow-green color, with small, speckled new leaves. Plants low in boron display brittle petioles, which tend to break halfway, leaving leaves dangling.

Tricks

Some growers like to plant caladium tubers upside down in order to get more green in the finished plants, though for most cultivars this practice is discouraged. Too many growers put all of their caladium crops in the same house under the same light levels. In reality, to grow excellent-quality caladiums, you need to experiment somewhat to find the best light level for your cultivar, as color pattern and intensity vary substantially. Truly well-grown caladiums are produced at just the right light levels.

When grown as bedding plants, varieties that tolerate full sun generally have at least one-third green coloration in the foliage. Store tubers at high humidity, but don't ever refrigerate them. Surflan (oryzalin) is a very good preemergent herbicide for this crop.

Interior Care

Protection from cold drafts is critical for this truly tropical foliage plant. Don't keep caladiums near doors or windows that open frequently. For best color and growth, give them as much light as you can, without exposing them to direct sunlight. Try not to let the temperature drop below 65° F (18° C).

Caladiums frequently die back and rest for a while. This usually happens during winter, and you can withhold water for some time to let tubers rest, then revive them in the spring with water and fertilizer. Caladiums can easily be damaged by cold irrigation water in northern climates, so let the water sit at room temperature before irrigating.

Lynn P. Griffith Jr. is president, A&L Southern Agricultural Laboratories, Pompano Beach, Florida. May 1998.

Calla Lily

Simple Steps for Pot Calla Lily Success

Peter Beckman and Tom Lukens

After years of popularity as a cut flower, calla lilies are now available as pot plants in many colors and plant forms available in the spring and summer-flowering "miniature" types.

The popularity of potted, colored callas has increased dramatically and has fueled breeding and research efforts. These advancements have helped growers be more successful with this somewhat tricky crop. Calla culture requires attention to detail. However, by following a few simple guidelines, most problems with calla bulb forcing can be overcome.

The calla lily, *Zantedeschia* species, isn't a true scaled bulb. It's a tuber or rhizome (native to temperate or subtropical South Africa) and its storage organ is more similar to a potato than a tulip. Although spring-colored callas are usually forced from November to June (as opposed to white winter cut aethiopica types), callas can be successfully grown in any season due to their photoperiodic neutrality.

Tuber availability is more likely to be limiting. Storage reserves from the tuber normally allow strong and vigorous growth for up to six months after harvest, and commercial research into controlled atmosphere storage will further widen the availability window. The primary limiting factor of calla production is a soft rot caused by the *Erwinia caratovera* bacterium. Once infection occurs, no chemical or other control can clean up or eradicate the bacterium from the tuber or potted plant. It's important to note that earlier invasive diseases such as pythium, phytophthora, or rhizoctonia generally stress or damage the calla and set up or even initiate the bacterial soft rot syndrome.

Therefore, cultural practices are largely focused on preventing these preliminary distresses. This is the best strategy for establishing a healthy, strong and desirable potted plant. Attention to temperature and water management, media selection, and a *preventative* disease program are essential.

Managing Disease with Environment

A well-drained soilless media with no more than 30 to 40% coarse peat (pH 6 to 6.5) allows a more aerated, warmer, and healthier root environment for forcing calla tubers in winter. Prolonged wet and especially cold soils encourage fungal infections that eventually lead to the soft rot syndrome.

Receiving and Planting Tubers

Unpack on arrival, and place in well-ventilated trays for three to seven days at 65° F (18° C) before planting. This allows abrasions to heal before planting. Dispose of any soft rot, and wash hands to avoid spreading pathogens. Long term storage should be done at 45° to 50° F (7° to 10° C) and in well-ventilated trays with a relative humidity of 75 to 90% to help avoid excessive desiccation.

Buy callas that have already been treated with bloom-enhancing gibberellins, or pretreat at 100 ppm Promalin and then air dry. Plant tubers with rounded side down and sprout side up (slightly tilted in large sizes) and between one to two inches deep. Water them in.

Fungicidal Drench

Within one to three days following the initial watering, drench with a combination of Subdue (2 oz. per 100 gal.) for water molds, plus a broad spectrum fungicide such as Fungo-Flo, Cleary's 3336, Chipco 26019, or Terraclor (label rates). This is absolutely essential for starting disease-free pots. Research has indicated some success in reducing soft rot with pre-plant bulb dips of copper or Streptomycin (200 ppm). Make sure active ingredients aren't depleted in dirty dip tanks, or pathogens will spread. Note: *Dips shouldn't replace drenches!*

Temperature and Light

A warm and fast start, especially during cool winter forcing, helps reduce bench time and disease incidence and reduces late or variable flower onset. Start pots at 70° to 75° F (21° to 24° C) (or 70° F/21° C bottom heat) until sprouts are well-emerged 2 to 3 in. After leaf unfurling, nights can be dropped to 60° to 65° F (16° to 18°C), and days can run 65° to 80° F (18° to 27° C). Warm temperatures will speed flowering, but will produce a softer plant. At flower initiation, cooler nights of 55° to 60° F (13° to 16° C) and high light (minimum 4,000 f.c.) conditions enhance flower coloration and reduce stem softness and elongation. Note: You can always slow a calla down with cool temperatures, but speeding it up with heat will soften plants and reduce keeping quality of the pot.

Fertilization

A balanced three- to four-month slow release (plus minors if possible) can be incorporated or top dressed. Calcium availability in media may help with disease management and flowering and is a good precaution. Liquid feed also performs well. Start at emergence, and go until first bloom with approximately 200 ppm nitrogen 20-10-20.

Watering

Water management is critical to calla health, especially in winter forcing. One thorough watering followed immediately by a fungicidal drench is often enough until sprouts are emerged. Then water sparingly until foliage is full.

Pots should neither remain constantly wet nor completely dry out. Avoid stress. Avoid pooling or splashing water pot to pot. The best success is often with weighted drip emitters, and the most trouble with disease is associated with ebb and flow type systems.

Plant Growth Regulators

Bonzi (paclobutrazol) drenches give the most effective height control and produce stouter, deeper-colored foliage. *Drench amount and timing is critical* and can be difficult due to variable locations, climates and foliage habit objectives.

Run in-house trials following these guidelines: Apply a concentration of ¼ to ½ oz. per gallon at 4 to 8 oz. per 6-in. pot when sprouts are 1 to 4 in. tall. This is often sufficient depending on your situation. Use the higher rate in lower light/shorter day conditions. An additional, similar application after two weeks may be appropriate if light conditions have been or are expected to be poor. (Bonzi is taken up by roots, and activity is greatly influenced by post-application respiration rates.)

Applications later than this are less effective, and earlier ones cause too much stunting. Uniform results are better when the drench is made about one day after watering and timed when calla sprouting among drenched pots is relatively equal. Subdue is compatible and may be appropriate for a second time with the Bonzi drench in certain circumstances, such as when pots are left particularly wet and cold or when current roots are slightly translucent or discolored.

Gibberellic Acid

Application is critical for good flower production, generally doubling counts versus untreated bulbs. Most bulbs, because most growers request it, are pretreated with gibberellic acid before leaving suppliers.

However, if you're doing in-house treatment, follow these formulas: Promalin and Provide (i.e. gibberellic acid$_{4,7}$) are best applied at 75 to 100 ppm and Progibb

(GA$_3$) at 125 ppm. Mix this with Physan-20 disinfectant at 1¼ tablespoon per gallon of gibberellic solution, and *spray* bulbs with a backpack on both sides. Allow them to air dry in trays without fans for approximately four hours.

Dipping bulbs can spread disease. Dip rates are 25% ppm less compared to the above. Do this for one to five minutes. Be sure Physan-20 is at 1,000 ppm or better using the manufacturer's test kit.

Insects

Generally speaking, forced calla pots suffer little from insect pest infestations. Fungus gnat control usually isn't a problem unless plants have been over-watered, or when soft rot is frequent. In this case, controlling flies is important to prevent any possibility of bacterial pathogen spread. Gnatrol, other biocontrols such as parasitic nematodes in Exhibit, or chemical drenches such as Diazinon at their label rates have proven effective. For the unusual thrips, mite, aphid, or whitefly infestations, label rates of the most common insecticides have been non-phytotoxic and effective.

Scheduling the Varieties

Forcing potted callas requires fifty-five to seventy-five days bench time. The number of days between planting and bloom decreases as the planting date moves later in the year and days are longer and warmer. Planting from November to mid-February, pinks and gems average fifty-five days to first flower and seventy days to the gibberellic acid-induced peak bloom, whites sixty and seventy-five days, and yellows or hybrids fifty to seventy-five days respectively. March to May plantings and larger bulb sizes of two inches and up generally require five days less bench time compared to the above.

Peter Beckman is plant pathologist and Tom Lukens is marketing manager-president, Golden State Bulb Growers, Watsonville, California. February 1997.

Chrysanthemum

Rewriting the Book on Garden Mum Production

Edward A. Higgins

Once upon a time, garden mum cuttings were planted in May, pinched two or three times, sold in September, and that was the end of the story. But new varieties, cultural techniques, and more sales opportunities are rewriting the book on garden mum production. In the process, garden mums are becoming a bestseller. We believe

sales have more than quintupled in the last decade.

Besides the traditional long-term crop, there are fast crops, two-cutting programs and early shaded programs. Even traditional crops are moving to early June planting instead of mid-May because of new cultural techniques and the production of new, faster growing and freer branching varieties, such as the Prophets, which now make up more than 87% of all garden mums sold.

Two-cutting Program

Instead of one cutting per pot and pinching each plant two to three times, plant two cuttings per pot and pinch each plant once. This eliminates up to two pinches—and the countless hours spent pinching—and up to forty days of production time. Instead of planting your natural-season fall crop in mid-May or early June with one cutting per pot, plant two cuttings per pot between July 1 and 7. Pinch this crop just once, approximately ten to fourteen days after planting. Fertilize and water well to encourage rapid growth. Flowering dates and plant size are comparable to crops planted earlier and pinched two to three times.

Fast Crop, No Pinch!

A fast crop of mums takes nine to ten weeks instead of the usual fifteen to sixteen. Because fast crop plants bloom about one week later than natural season crops, they offer another way to extend the garden mum season. Finished fast crop plants in 6-in. pots are smaller—10 to 14 in. in diameter—than regular garden mums because of the shortened growing season. This allows growing nearly twice as many plants in the same area. These production efficiencies should enable growers to stimulate sales by marketing fast-crop garden mums at a different price point than regular crops.

Plant 6-in. pots and all pan sizes between July 16 and 30 and 4-in. pots between August 1 and 16. No B-Nine is required and *don't pinch*. For 8-in. pots, first follow this schedule, but plant two cuttings per pot. Remember, don't pinch! The naturally occurring changes in day length will cause a crown bud to do the "pinching" for you.

Early Shaded Programs

Many garden centers want flowered garden mums in July and August to create an extended sales period from the end of the bedding plant season until fall plants are in the market. Planting times for these programs begin the first week of April and end June 5. Generally, one cutting is planted in a 6-in. pot and pinched twice. Short days (blackcloth) are provided to schedule the crops to bloom for specific dates.

Regardless of which program you choose, the key elements of successful garden mum production remain the same: Plant promptly, use well-drained media, keep media moist, keep plants well-fertilized, use the best possible cultivar, and provide adequate spacing. Don't stress garden mums early in the crop, or branching action may be reduced and premature budding encouraged.

Planting

To get plants off to a vigorous start, plant cuttings immediately in a moist medium. Water in right after planting using a 200 to 300 ppm NPK fertilizer. Mist or syringe plants frequently for the first few days or until they are fully turgid with the roots actively taking up water.

Soil Media

Media should be loose and well-drained. The pH for soil-based media should be 6 to 6.5 and between 5.5 to 6 for soilless media.

Containers

While a variety of container sizes are used in garden mum production, the most popular for summer and fall are 6-in. to 8-in. pots and 1-gal., 1.5-gal., and 2-gal. nursery containers. In spring, cell packs and 3-in. to 6-in. pots are widely used.

Spacing

Beautiful, mounded plants develop with adequate spacing, while those grown too close together can have a stovepipe look. The desired finished quality and selling price play important roles in determining the distance to use. Six-in. pots for spring, summer, fall, and fast crops are usually spaced on 12- to 15-in. centers. Eight-in. pots and larger for fall are spaced on 18- to 24-in. centers.

Fertilization

Fertilization rates vary, depending on the medium, fertilizer, and also on the frequency of application. Generally, a 250 ppm N constant liquid feed produces high-quality garden mums. The rate may need to be adjusted, depending on the soil mix and weather. Balanced fertilizers such as 20-20-20 and 15-15-15 are suitable. Continue fertilization until buds approach pea size. If liquid feeding is not possible, slow-release fertilizers may be used.

Watering

Plants should never be stressed in the first half of the crop! Foliage, however, should be dry before evening. Drip or tube irrigation is preferred to overhead watering to help bacterial leaf spot, botrytis, septoria, alternaria and other foliar diseases.

Pinching

Pinch when plants have 1 to 1½ in. of new growth, not by calendar dates. This usually occurs ten to fourteen days after planting, and the top ½ in. of new growth should be pinched off. Some growers find it economical to plant three to five weeks later than normal using two cuttings per pot and pinching only once.

Insects

Aphids, mites, caterpillars, leafminers, and thrips may attack garden mums. Fortunately, insects usually are not a significant problem.

Diseases

The most common diseases are root rots pythium and rhizoctonia and leaf spots caused by botrytis or bacterial leaf spot. In contaminated media or field soil, fusarium may also develop.

Edward A. Higgins is chrysanthemum product manager, Yoder Brothers Inc., Barberton, Ohio. June 1995.

Florel Shows Great Promise When Used Properly

Florel, a new growth regulator, may be the greatest growing tool since water, but it's not a crutch or Band-Aid for poor culture. You can't spray a crop that is already poorly budded and expect magic.

While more trials remain to be run, Florel shows good potential to delay fall flowering response, reduce or eliminate pinching, and inhibit or reduce the number of early crown buds. Florel promises to be an important tool in managing garden mum growth.

First application

For best results, mix according to the labeled rate of 500 ppm. Apply only after plants are well established—five to seven days for rooted cuttings and ten to fourteen days for unrooted cuttings. Never apply to plants that are wilting, fatigued, or otherwise stressed.

How to apply

For best results, use a spreader sticker in spray solutions with deionized water or water with low alkalinity. High alkalinity can diminish Florel's effectiveness.

Pinching

Manually pinch when the cuttings are ready for their first pinch. A word to the wise: Set aside a few plants that you won't pinch; just spray the plants with Florel. Use them to help determine whether to pinch at all the next year for the first pinch or to just use Florel instead.

Second application

This should come approximately two weeks after pinching, again using a 500 ppm solution. If made by late June to approximately July 5, you can stop applications and experience little or no delay in flowering, yet still obtain a bushier, fuller plant with more uniform bud formation than plants grown without Florel.

Subsequent applications

If the goal is to delay flowering, continue Florel applications every two weeks until early August. This will delay flowering of some early and midseason varieties into late September and October. However, there are as yet no accurate guidelines to use in predicting flowering times by cultivar based on the number of Florel applications.

Culture Notes, June 1995.

Producing Quality Winter Pot Mums

Winter pot mums, planted in late October and November to flower in January and February, are among the most challenging mum crops to produce quality plants that finish on time. Here are some tips from Yoder Brothers Inc. to keep your winter pot mum production smooth this season.

Carbon Dioxide

The ideal injection of carbon dioxide is 500 to 1,000 ppm. It's particularly important during winter when ventilation is reduced. Also, when HID lights are used, carbon dioxide must equal the light intensity, or photosynthesis could be limited.

Height Control

Because days are cloudy and cooler and day and night temperatures aren't as far apart, plants are naturally shorter because internodes don't stretch as much and growth is slower. Yoder recommends delaying your first B-Nine application (don't exceed 2,500 ppm during this period) until new shoots are 3 to 3½ in. long. Short varieties may not need growth regulators.

Light

One of the factors that most affects winter pot mum quality is low light. To keep plants in adequate light, make sure greenhouse roofs are clean, and don't hang crops above your mum crop. Also, keep spacing adequate, especially early in the crop when tight spacing can reduce the number of breaks after pinch. Use HID lights at 500 to 800 f.c. for eighteen hours daily in the North until two weeks after starting short days (up to eleven hours daily). Note: If your winter crops were too short last season, try adding five to seven long days after pinching. This can improve quality, especially on short treatment varieties.

Nitrate-Nitrogen

Provide nitrogen in the nitrate form. Also, fertilization is less frequent in winter, so increase your fertilizer rates 50 to 75 ppm. If you're mixing your own fertilizer, switch to calcium nitrate and potassium nitrate until April 15. During winter, Yoder suggests complete fertilizers with 50% or more nitrate.

Pinching

Before pinching, make sure plants are ready—look for 1 to 1½ in. of new growth. Also, roots should reach the bottom and sides of pots. Pinching too early can cause plants to lack fullness, as they will develop a limited number of lateral branches.

Temperatures, Timing

Minimum temperature (without risk of delay) should be 65° F from the time from planting until visible bud, when temperature most affects crop timing. From visible bud to disbudding, lower night temperatures to 63° F; from disbudding to flowering, lower night temperatures to 60° F. Keep day temperatures on cloudy days only 0° to 5° F warmer—high temperatures without high light intensities weaken plant growth. Note: Raising temperatures after visible bud stage *will not* speed up your mum crop.

Keeping Quality

You can significantly improve keeping quality for consumers by providing the maximum available light and fertilizer two to four weeks before finish. Flowers should be open 50% or more before sale. Avoid leaf yellowing and decreased postharvest life by shipping and storing plants at 35° to 40° F and keeping shipping time under seven days.

Culture Notes, November 1997.

Hefty Hardy Mums

Chris Beytes

Jan Barendse, Baker Greenhouses, Utica, New York, believes in growing, and profiting from hardy mums. No 4-in. or 6-in. pots for him. Rather than cut costs with

smaller pots and fewer cuttings, Jan puts four cuttings into a 10-in. terra-cotta-colored planter that wholesales for $7 to $8. Here's how:

- Stick unrooted cuttings into fifty-cell trays in the middle of May.
- Pot in early June. He uses six- and seven-week varieties, for a sales window from August 20 to mid-September.
- Brewery sludge (hops and barley hulls) helps hold moisture and gives his soil mix some structure.
- For fertilizer, Nutricote 14-14-14. He also gives a liquid feed of calcium or potassium nitrate after a heavy rain.
- Pinch once, ten days after potting.
- No growth regulators.
- Irrigation is overhead.

"If the quality is good, people will buy," Jan says. He must be right; Baker Greenhouses sold 10,000 in 1994 to garden centers and chains and is planning to grow 15,000 to 20,000 in 1995.

Chris Beytes, editor, GrowerTalks *magazine, Batavia, Illinois. June 1995.*

Dip Cuttings in B-9 before Planting
for More Compact Mums

Applying B-Nine to chrysanthemum cuttings as a preplanting dip reduces the height at which the pinch is made, resulting in more compact plants, Jim Barrett, University of Florida, said at a recent [1995] growth regulator meeting in England. This could also be promising for hardy varieties with long internodes. Growers in the United States, where B-Nine is widely used, have already been tri-aling the technique. B-Nine is also used on bedding plants, pot begonias, and many nursery stock plants. Unlike other growth retardants, it doesn't cause uneven growth or leaf spotting. Because B-Nine is less active than other chemicals, researchers are also searching for ways to increase its activity. Spraying the chemi-cal before sunrise or after sunset gave 25% more growth reduction than sprays applied during the middle of the day, as the spray can only penetrate leaves while it's still wet on the leaf surface. They also found that Cell-U-Wet, a new spray additive, delayed the spray's drying—and increases its activity.

Culture Notes, September 1995.

Coleus

Coleus for Sun *and* Cold

George Griffith

The Solar series, a group of coleus from Hatchett Creek Farms, Gainesville, Florida, can be grown in almost any climate, thriving in 68° F (20° C) night temperatures and in full sun. The Solar series offers an excellent growth habit and is self-branching. The varieties don't require growth regulators and, most important, they don't flower freely. The series was trialed at the University of Georgia, Athens.

Tolerating a wide range of growing conditions, the Solar series is more cold tolerant than any other coleus known thus far. Color retention is quite good, showing more intense coloration when grown under cooler night temperatures (below 68° F/20° C), with most of the series developing a metallic sheen. Plants will change color when grown under high temperatures but won't fade. Lower fertility levels, especially a lack of nitrogen, will increase color intensity.

Growth habit and size are mainly regulated by giving adequate amounts of water and fertilizer. Higher levels of nitrogen will cause succulent growth and formation of larger leaves. Low nutrition levels, nitrogen in particular, will result in more intense color and smaller leaves but will induce more flowers.

The Solar series is almost maintenance-free during the growing season and can be grown under various light intensities from medium shade levels (3,000 f.c.) to full sun (10,000+ f.c.). The Solar tone, however, will be greener when grown under shade. The new series doesn't originate from hybrid cultivars, and most of their seeds aren't available. Propagation should only be through vegetative means from tip cuttings.

Originating in the tropical mountains and lowlands of Indonesia, the series demonstrates a wide tolerance to varied soil conditions under tropical conditions. It grows equally well in the lowlands (164 ft. above sea level, with night temperatures averaging 75° to 86°F/24° to 30° C) and in the highlands (2,789 ft., night temperatures averaging 64° to 68°F/18° to 20° C). Color intensity may change under lower night temperatures—pink and dark pink colors turn bright red with a bluish, metallic sheen.

With adequate fertilization, the Solar series isn't a prolific bloomer, meaning less maintenance. You don't have to remove flower buds once they become visible to keep plants looking attractive. Solar series varieties don't need to be grown under partial shade and are quite long lasting, up to six months. The series includes Solar Sunrise, Solar Shadow, Solar Storm, Solar Eclipse, Solar Flair, and Solar Shade.

Hurricane Series

The Hurricane series, a group presenting unique, fan-shaped, twisted, and fluted leaves, is the newest coleus from Hatchett Creek. The three varieties from this series, Hurricane Susan, Hurricane Louise, and Hurricane Jenni, have similar growth habits and attributes to the Solar series.

Hurricane Susan has a predominantly yellow color with green lines at the center and is the shortest of the three. It likes full sun and is rain tolerant. Susan is a heavy feeder and has very few flowers. It's resistant to aphids and chewing insects such as grasshoppers. It offers an excellent, upright growth habit, growing to approximately 1½ ft. in height under poor soil conditions and up to 2½ ft. with high fertility. Susan is excellent for backgrounds and beds. Sporadic flowering occurs under tropical conditions (twelve hours of daylight).

Hurricane Jenni is the most colorful of the three and grows slightly taller than Hurricane Susan. It presents a kaleidoscope of colors including bronze, maroon, yellow, and dark and light green.

Hurricane Louise is the tallest of the three, with multicolored orange and bronze leaves and a dark and light green center. Like the Solar series, its color intensity varies depending on night temperature and fertility level.

George Griffith, Hatchett Creek Farms, Gainesville, Florida. February 1996.

Cyclamen

❋

Growing the Best Cyclamen You Can

Teresa Aimone

Cyclamen have gotten a bad rap over the past few years. Their long crop time (particularly when grown from seed) and rather exacting growing needs can be frustrating and turn growers toward easier, less demanding crops. Growing this

crop brings definite rewards: Cyclamen command more money, both wholesale and retail; they can be finished from November to May (this includes December 26, when those last few poinsettias may not be selling so well).

Cyclamen don't like stress, and they don't like abrupt changes to their environment. Cleanliness and consistency are the keys to growing this crop. Rapid fluctuations in temperature, fertilization, and watering regimes; constantly moving plants from one place to another; and unsanitary growing conditions make this crop a lot harder to grow than it needs to be. Keep careful records when growing so you can repeat what went right. Remember: Cyclamen grow slowly, and changes you make in production today may not show up for two to three months in a cyclamen crop.

Germination

When germinating cyclamen (or any seed), it's important to look at the seed under a hand lens or microscope to see exactly what you're starting with. In the case of cyclamen, you'll see irregularly shaped seed that varies in color from light tan to dark brown or almost black. The seed coat is *very* hard, and it takes three weeks or more for the seed coat to soften and the radicle and cotyledon to emerge.

Media

The main ingredient in the germination media should be peat; this will help you maintain high humidity around the seed. Some growers use straight peat to germinate cyclamen. If you want another component in your mix, add perlite in a 3:1 peat to perlite ratio. Perlite will provide good aeration in the media. Avoid using just vermiculite as your second component, as vermiculite compacts and can make the media too dense.

Container

Cyclamen can be germinated in open flats or in a plug tray such as a 128 or larger. If using an open flat, sow seed on 1-in. by 1-in. spacings. Using a larger sized plug makes transplanting easier and also avoids an interim transplanting stage.

Covering

Cover the seed lightly with a coarse grade vermiculite. This is to keep moisture around the seed and allow some air movement, too. Remember what's happening at the seed level!

Light

Germinate the seed in *total* darkness. The easiest way to do this is to turn the lights off in the germination chamber. Another way is to invert a clean flat over the germinating flat. Or some growers put germinating flats on carts and wrap the whole cart in black plastic.

Moisture

You can either moisten the flats slightly after sowing or moisten the media before sowing. The key is to maintain as close to 100% humidity as possible during germination. Remember, the seed coat is very hard, and you want to soften it up as much as possible. Check the humidity level every three to five days. Use a mist nozzle to raise the humidity if it's too low. The media right around the seed should feel moist—kind of like a damp sponge.

Providing high humidity will help you avoid "sticks." A stick is what happens when the cotyledon (there's only one) is completely enclosed inside the seed coat. If the seed coat is stuck to the edge of the cotyledon leaf, that's not a problem; we're talking about the seed coat completely enclosing the cotyledon. If you see this happening, it's a definite sign that your humidity is too low. Some growers have experienced cyclamen seed sticks on up to 30% of their crop. If you don't get the seed coat off, that plant will never grow, and you can just throw that seedling away. To remove the seed coat, you must first soften it up by raising the general humidity or by directly misting the sticks. The seed coat may fall off, or you may have to physically remove the seed coat by hand. It sounds time consuming, but your alternative is to throw that plant away.

Constant temperature

Maintain a 65° to 68° F (18° to 20°C) constant soil temperature. If you can't maintain constant temperatures, try to stay within a 3° F window. Stay above 60° F (16° C) and below 72° F (22° C). Whatever temperature you're having success with, the key is not to change it.

pH

Maintain a 5.5 to 6.3 pH toward the end of the crop time. You can let pH go up to the higher end of this range, but for germination, stay around 5.5 to 5.8.

Germination time

Germination time is three to four weeks. After germination, move the plants to a shaded part of the greenhouse (30% shade is fine) to avoid leaf scald on the cotyledon and emerging true leaves.

Disease control

If needed, you can use a fungicide drench on cyclamen at the time plants start to get their first true leaf. This occurs about two to three weeks after taking plants out of the germination area.

Growing On

Cyclamen need to be transplanted into an interim container size such as a 72- or 55-cell plug only if you have germinated in a small plug, such as a 288, or if you're germinating in open flats. Basically, you transplant when leaves start to touch—about eight to ten weeks after sowing. Grade plants as you transplant into the interim container to make watering easier. If you've already started plants in a larger cell, you don't need to transplant. Plants will stay in these containers or in the original larger plugs you germinated in for another ten to fourteen weeks. Plants will have four to six true leaves on them when you transplant into the final container.

Planting depth

Cyclamen will have started to form a corm at this point. The corm is the round, bulbous structure from which stems, leaves and roots arise. Botanically, a corm is a compressed stem. For planting, imagine a line drawn around the widest part of the corm; you don't want to plant above or below that line. Planting above the line can cause the stem to settle in too deeply and rot. If you plant too high, the corm can dry out, you'll have poor root development, and the plant may be wobbly and unstable. Plant the corm flush with the top of the media. It will settle into the media when you water.

Disease protection

You can drench plants again about ten weeks after sowing or when two true leaves have fully developed.

Final containers

Transplant into final containers when plants have four to six true leaves. This is the same stage as you would buy liners. Plants can be grown pot tight and will remain in that stage for about another eight weeks or until leaves start to touch for final spacing.

Temperature

Because seed is normally sown in February or March for Thanksgiving or Christmas flowering, production in final containers occurs at the hottest time of the year. This is the moment of truth for cyclamen. Both purchased and home-grown plants will have been happily settling into their routine for about the past twenty weeks, and that is the time when they're stressed by transplanting and cool temperature and they become most critical and hardest to maintain.

Cyclamen types are divided according to flower size:

Type	Flower size	Varieties	Finishing container
Miniature	1 in. or less	Dixie, Miracle	4 in. or smaller
Intermediate	1 in. to 1½ in	Laser, Junior	5 in.
Standard	2 in. and larger	Rondo	5 in. to 5½ in.
			Concerto 5½ in. and larger
			Pannevis 6 in. and larger

For the first seven weeks after potting maintain 65° to 68° F (18° to 20° C) night temperatures to promote root development. For the next seven weeks maintain 62° to 65° F (17° to 18° C) temperatures. Day temperatures can be 5° to 10° F higher. How do you maintain these temperatures? With fan and pad cooling, sectioning off greenhouse space, shading, and above all, air movement at the plant level. Simply keeping air flowing through the dense canopy of cyclamen leaves can be a big help in growing cyclamen through this tough time of the year.

Watering

Cyclamen do well with any kind of watering, such as ebb and flood, mat, or tube watering. Overhead watering is difficult to do effectively with cyclamen because they have dense canopies, and the cyclamen leaf shape helps funnel water away from the center of the plant. Keep water off the leaves, and avoid watering late in the day unless plants are so dry they'll collapse by morning.

Fertilization

Cyclamen aren't heavy feeders. Avoid ammonium forms of nitrogen; instead, use calcium or potassium nitrate as your nitrogen source. The first feedings are in the plug stage at 50 to 75 ppm; feed every two to three weeks. Increase feed gradually by 25 ppm every month until you top out at 150 to 200 ppm. Leach with clear water every three to four waterings. Regular soil testing will help you closely monitor nutrition. Remember, if you're using bottom irrigation, salt levels will be higher at the top part of the pot, so keep that in mind when you're taking your soil samples.

Alternate your feedings with a more potassium-pronounced feed to encourage root growth. Use a ratio of 1:2 or 1:1.5 nitrogen to potassium. This is especially true if you're growing F_1s. Open-pollinated plants require more nitrogen instead of potassium.

About three weeks before shipment, increase phosphorous levels to help tone plants, brighten flowers, and increase shelf life. Some suggested sources of fertilizer are 9-45-15 or 5-50-17.

Light

Cyclamen require shading during hot summer months to avoid leaf scald. Northern climates can use 30% shade; you may need up to 50% in more light-intense southern regions. Petioles will stretch if shading is too intense, so adjust shade accordingly. A good rule of thumb to follow regarding time to shade: When trees leaf out in late spring/early summer, put shade on. When leaves fall off trees in fall, take shade off.

Diseases and Insects

With cyclamen, you need to adopt a "take no prisoners" attitude. Sanitation is the best disease prevention. Keep walkways and bench tops clear of debris. Don't leave the water breaker on the ground; hang it up off of the floor. Use plastic liners in garbage cans, and use a new liner after taking out the garbage. Don't throw dead or diseased plants away if you don't remove potential disease sources by taking them all the way out of greenhouses. Employees should wash hands before handling crops. When removing spent blooms and leaves, the entire petiole or peduncle needs to come off. Grasp the bottom end of the peduncle or petiole at the point where it attaches to the corm, give it one or two turns clockwise and pull it off. Leaving part of it behind is an excellent entry point for disease.

Fusarium

Symptoms include rapid yellowing, wilting, and death, often on one side of the plant. The corm will have brown or reddish streaking, along with a purplish discoloration. The leaf petiole can also be pinched or blackened. Remove and destroy infected plants.

Erwinia soft rot

Also called bacterial soft rot, this disease is evidenced by sudden wilting, followed by plant collapse. Corms will be soft and slimy and have a strong, unpleasant odor to them. Remove and destroy infected plants.

TSWV

Symptoms start as thumbprint or circular yellow ring spots on leaves. Symptoms may include brown streaks on leaf petioles; you can also get misshapen corms with brown streaks inside. Flowers may also be streaked and malformed (often a sign of the disease's vector infestation—thrips). Roots will usually look fine until plants are about to die. It can take three months from the time of infestation for symptoms to appear. Destroy infected plants. Monitor thrips populations and use appropriate chemical controls.

Cyclamen mites

Infestations are centered around the plant corm. The most common symptoms include reddish/yellowish stippling or discoloration to leaves; leaves can also get rather stiff or become curled and distorted. You can also have twisted flower stems or flowering under the foliage.

Teresa Aimone was a regional specialist Southeast, S&G Seeds, Coppell, Texas. October 1996.

Daffodils (Narcissus)

❈

Pot Daffodils

Teresa Aimone

The following information is for daffodil bulbs potted at the end of August and forced for Christmas/early January sales. Precool bulbs before planting. Beginning the last week of August, store bulbs in a dark, well-ventilated cooler, keeping the cooler ethylene-free to prevent premature forcing. During the first week of October, plant bulbs in an azalea pot or a bulb pan. Media pH should be 6.0 to 7.0. Planting suggestions: three bulbs to a 6-in. or five to six bulbs to an 8-in. container. Daffodils can produce enough roots to push themselves out of the pot, so be certain to plant bulbs deeply enough—just so the tip of the bulb is showing. Water bulbs in after planting, and routinely check bulbs for watering throughout cold storage. Also, check for root development for four to six weeks after planting.

For most daffodil varieties, optimal cold storage time is fifteen to sixteen weeks; the production schedule for Christmas/early winter requires less time. Maintain 49° F (9° C) until November 5 to 10, then drop the temperature to 41° F (5° C) until plants are brought into the greenhouse. For flowering at Christmas, bring plants into the greenhouse on December 1. The first watering should be very thorough, then keep soil moist after that. Plants will bloom approximately three weeks after they're brought out of the cooler into the greenhouse for forcing. Highest quality plants result from forcing at 60° F (16° C) nights and 65° F (18° C) days. Spray plants with ethephon (Florel) if height control is necessary. Two-and-a-half gal. of solution will cover five hundred 6-in. pots. *The Holland Bulb Forcer's Guide* recommends the following when spraying with ethephon:

- Leaves and/or floral stalk should be three to four inches long.
- Foliage must be dry at the time of treatment; late afternoon is preferred after a morning watering (if watering is needed).
- Don't wet foliage for twelve hours after treatment.
- If a second application (2,000 ppm) is used, apply it two to three days after the initial spray. Don't apply if the flower bud is highly visible in the foliage, or flower abortion may occur.
- Apply in a well-ventilated 60° to 63° F (16° to 17° C) greenhouse.

Teresa Aimone was a regional specialist Southeast, S&G Seeds, Coppell, Texas. August 1997.

Delphinium

Success with Perennial Cuts: Delphinium

Jeff McGrew

The delphinium group is a large and important one. *Delphinium consolida* is an annual delphinium commonly called larkspur delphinium and is one of the top performers for both the fresh cut and dried flower industries.

Flower Types

The so-called delphinium hybrid group (*Delphinium* x *cultorum)* contains the large

double-flowered types (mostly hardy perennials) called the Giant Pacific hybrids. These cultivars account for 30 to 40% of the delphinium cut flower market.

The *Delphinium belladonna* types are also perennial, and they account for about 6% of the delphinium market. Belladonna types are large, single flowers arranged symmetrically on 12- to 16-in. stems. Colors can range from dark blue (sometimes called *D. bellamosum*), light blue (*D. belladonna* or Clivedon Beauty), and a pure white (called *D. Casablanca*). Of the belladonna types, the dark blue varieties have a larger market share than the whites or light blues.

In general, delphiniums appreciate cooler growing temperatures—nights 46° to 54° F (8° to 12° C) and days 59° to 77° F (15° to 25° C). Delphiniums can be profitably grown in open fields or greenhouses.

Greenhouse Production

For greenhouse forcing, transplant a seven- to ten-week-old plug into a cool greenhouse between October and March in the Northern Hemisphere. Maintain cool night temperatures (46° to 54° F/8° to 12° C) during the early production period (first six to eight weeks after transplanting). This will maximize flowering potential.

Two support wires are recommended. Space plants on 10-in. centers. Drip irrigation, not overhead watering, is recommended.

Total crop time ranges from ten to fourteen weeks (after transplant), depending on planting time and production temperature. If plants are properly grown, you can achieve two to three flower flushes in one season.

Outdoor Production

Outdoor field growing is the most common production method. Transplants are put out in August through October or March to April in the Northern Hemisphere. In cold growing areas, protect young transplants from freezing temperatures with row cover cloth.

Space plants 8 to10 in. apart. Support wire is also recommended for outside production along with drip or bottom irrigation.

In northern growing environments, one to two flower flushes are normal. In a more moderate growing area, two to three flushes are possible.

Postharvest

Harvest flower stems when 20 to 30% of individual flowers are open. Use a bactericide and an ethylene inhibitor, such as silver thiosulphate (STS), to prevent petal shattering and increase vase life.

Jeff McGrew, director, North America, Kieft Bloemzaden B.V., Mount Vernon, Washington. December 1997.

Exacum

Growing Exacum Your Way

Paul R. Cummiskey

Few plants are as versatile as exacum. Use exacum as an accent decoration item in the modern home or in windowsill culture, or try it outside in patio bowls and baskets or as a mass planting in semi-shady conditions. Combination plantings using all three colors—blue, white, and rose—together make a very striking arrangement.

For the grower, exacum can be grown all year and scheduled to flower during certain holidays with little cultural manipulation. Plants can be grown in many different size pots from 4 to 24 in., depending on the cultivar. Boxing and shipping exacum over long distances is easy compared to many other pot crops, with no

special requirements other than maintaining temperatures in the 50° to 70° F (10° to 21° C) range.

History

Exacum affine is a member of the Gentianaceae plant family. The genus is composed of sixty-five species from Africa, Saudi Arabia, Socotra, India, Southeast Asia, and New Guinea. Most species have remained relatively obscure to cultivation, with the exception of *Exacum affine,* which was collected on the island of Socotra off the southern Saudi Arabian coast by an English expedition in 1880. Exacum soon became a popular greenhouse pot plant in England, and all cultivars in use today have been developed from this species.

During the past ten to fifteen years, substantial improvements have been made over the original *E. affine* Midget. Today cultivars are bred for 4-in., 5-in., and 6-in. production. Some cultivars perform best indoors, while others perform very well outside in beds from 40° to 90° F (4° to 32° C). Breeders are presently working to improve garden performance, increase flower size, improve color selection from dark blue to violet, and deep rose to light pink, and improve the performance of double-flowering types.

Propagation

Exacum seed can be broadcast in open seed trays or sown by machine into plug trays. Sow seeds on the growing media surface or on top of a thin layer of vermiculite sprinkled on the growing media surface. Keep seeds evenly moist until seedlings are well established (four weeks). Four to five weeks after sowing, plug trays or seed tray should dry before watering to promote a good root system and prevent the development of any diseases, especially stem canker botrytis. Fertilize with 50 to 100 ppm of nitrogen every other watering or 250 ppm every third watering during this production period. In six to eight weeks, seedlings are ready to transplant into 1-in. or 2-in. liners or into the desired finished pot size. Spray with a fungicide for botrytis control at each transplanting (see Pest and Disease Control). For vegetative propagation, exacum must first be sprayed with a botrytis-control fungicide two to three days before cuttings are taken. Take two node cuttings from plants that haven't developed any flowers and remove any visible buds. Place cuttings on a misting bench, and use a very fine mist head to keep plants turgid but not overly moist. Allow four weeks for rooting at 70° F (18° C). Exacum can be propagated easily by tissue culture, but this method is costly and the final product is of equal or lesser quality than plants produced from seed production. Most growers purchase plugs from specialty producers.

Production

Crop timing is seasonably variable with a marketable 6-in. flowering plant being produced in seven to eight weeks in the summer and up to twelve to fourteen weeks in the winter from a 2-in. liner. Smaller plants in 4-in. pots can be produced in less time depending on the cultivar being used.

Temperatures during the finish production stage should range from 60° to 65° F (16° to 18° C) at night and 72° to 78° F (22° to 26° C) daytime. Plants can be grown at higher daytime temperatures to decrease production time and increase vegetative growth.

Light Requirements

Grow plants under full sunlight during winter months. In summer apply a light shade of 3,500 to 4,000 f.c. During spring and early fall a very light shade of 4,500 to 5,500 f.c. should be applied, especially when plants are starting to flower. Excessive light and heat will cause flowers to fade quickly and plants to set buds and bloom too early.

Exacum flowers in response to total light energy received on leaves. Flower bud initiation is not affected by daylength, but plants will flower more uniformly and earlier under long day conditions. Using supplemental lighting during winter is beneficial. Lighting such as HID lights or even mum-type lighting of 10 to 20 f.c. used four to six hours a night from dusk can decrease production time by two weeks during winter to early spring.

Growth Regulators and Height Control

Use growth regulators such as B-Nine or Bonzi on exacum to control height and produce a higher quality plant. From fall through spring, one application of B-Nine at 2,500 ppm increases plant quality, producing a better branched and shaped plant. Some cultivars are very dwarf and may not need an application of B-Nine during any production period. Consult growing information on the cultivar you're using. More than two applications of B-Nine will delay flowering, especially during winter months. Bonzi can be used on exacum with good results and causes very little delay in flowering. Use it at 0.5 to 1.0 ppm as a drench or 25 to 30 ppm as a spray.

Fertilization

Exacum responds best to moderate fertilizer levels. Use 250 to 350 ppm every second or third watering and higher rates during summer months. Constant feeding of 150 to 225 ppm can also be used. Exacum prefers higher levels of calcium and copper in the leaf tissue. Use a 20-10-20 or a 15-16-17, and alternate one of these fertilizers with a Hi-Cal fertilizer or use calcium nitrate plus potassium nitrate for higher calcium levels. A foliar spray of Tribasic copper or Phyton 27 twice during the early production stage (three weeks after potting) will help produce a healthier looking plant. During late spring and summer top-dressing with a slow-release fertilizer is beneficial, especially when liquid fertilizers are leached out very quickly during high temperatures. If fertilizer levels are too low, plants will flower at too young a stage, producing inferior plants.

Pest and Disease Control

The most prevalent disease problem on exacum is stem canker botrytis, especially during winter months, although fusarium and TSWV can also affect exacum.

Stem canker botrytis on exacum causes a gray to tan lesion on the stem either at the soil line or slightly above or below the soil line. Most infection in young seedlings occurs within one to two weeks after potting. Some plants may die immediately, while others may die one or even two months later. For disease prevention, it's very important to treat plants before they become infected. Treat young transplants with Chipco 26019 at 1 lb. per 100 gal. Cover the entire foliage of the seedling, and apply enough liquid to penetrate ½ in. to 1 in. of the soil. Waiting two to three days to treat exacum will mean an increase of stem canker botrytis. Spray plants a second time in four to five weeks.

Other important steps to take for disease prevention are to disinfect benches, maintain good horizontal air circulation, provide lower nutrients during winter, keep fungus gnat populations at low levels, and avoid overhead watering.

Broad mites can cause the most damage to exacum. Usually found on the upper parts of the plant, broad mites cause leaf and growing tips to become yellow and distorted, and buds will fail to open. Use Avid 0.15EC for controlling broad mite.

Thrips can also attack exacum and can carry TSWV, which can kill exacum in only a few days. The extremely small, highly mobile insects also cause physical damage to plants. Be sure to keep thrips populations low by monitoring populations and spraying at least once a week.

Paul R. Cummiskey, vice president research and development, Earl J. Small Growers Inc., Pinellas Park, Florida. October 1995.

Geranium

———————————— ✸ ————————————

Top Ten Tips for Perfect
Ivy Geranium Baskets and Pots

Karl Trellinger

Producing perfect ivy geraniums is highly desirable and financially can be very reward-ing. Many newer varieties are better branching with earlier flowering and more flower power. They have more colors and are less prone to edema. This makes growing ivy geraniums baskets easier and makes the product more appealing to the customer. By following these top ten tips, both retailers and growers can boast the perfect ivy!

1. Variety Selection

The main characteristics for great performing ivy geranium varieties in pots and baskets:

- excellent branching for maximum flower power and ease of transportation
- earliness for good overall appearance at time of sale (50% of flowers open)
- bold, clear colors
- edema resistance
- beautiful leaves
- resistance to shattering.

2. Timing and Amount of Cuttings per Pot

The most common mistakes many growers make:

- not enough cuttings per pot
- not enough plants in the center of the pot, which results in openness
- not enough height control
- not enough growing time when fewer cuttings are used
- not enough attention to the first—and most important—part of the crop.

Ten-in. baskets should have four to five cuttings, with one to two cuttings in the center and three cuttings 2½ inches from the edge of the pot. Start specimen baskets with rooted cuttings beginning in February. Start compact baskets for the mass mar-ket at the end of February for May sales. With three cuttings per pot, plants should generally be potted four weeks earlier. For high-quality plants and better price at retail, we recommend using seven cuttings in a 12-in. pot. Cuttings should gener-ally be arranged with four around the outside, approximately 2½ inches from the edge of the pot, and three cuttings in the center.

66

The market is still wide open for 4- to 5-in. ivies planted the middle of March (with Florel, not pinched) and for 6- to 7-in. pots with two cuttings per pot planted in the beginning of February for use in large containers and window boxes and as ground covers. With this high-density cropping, you can win new customers by being able to offer new product forms at more competitive prices that can also be planted together with other bedding plants.

3. Medium, pH, Fertilization, Feeding Schedule

Medium

The ideal medium has a pH of 5.1 to 5.6, 25% perlite, good coarse peat, and 10 to 15% vermiculite or clay for better water retention and buffer capacity.

pH

To achieve a pH of 5.3 (zonals prefer 5.8), which is ideal for ivies, test soil for pH and EC every two weeks. Make adjustments by choosing appropriate basic (calcium nitrate) or acidic fertilizers or sulfuric acid.

Fertilization

Conduct complete soil analysis at the beginning of the crop, then every four weeks after. In the beginning the analysis should read as follows: 150 ppm nitrogen, 20 ppm phosphorus, 200 ppm potassium, 100 ppm calcium, 50 ppm magnesium, 2 ppm iron, 1 ppm manganese, 0.3 ppm zinc, 0.15 ppm copper, 0.3 ppm boron, 0.05 ppm molybdenum. Six weeks after planting, until the end of the crop: 220 ppm nitrogen, 30 ppm phosphorus, 300 ppm potassium, 150 ppm calcium, 80 ppm magnesium, 3 ppm iron, 1.5 ppm manganese, 0.4 ppm zinc, 0.2 ppm copper, 0.4 ppm boron, 0.06 ppm molybdenum.

Feed so that 15 to 20% of the applied solution drips out of the pot to avoid salt buildup. On flood benches, the amount of fertilizer used should be normal the first four weeks, then cut in half. The soil has to be tested every two weeks. The following fertilizer can be used at 250 to 300 ppm constant feed:

- 15-15-15 and 20-10-20 for softer growth and lowering pH
- 15-10-30 for harder growth and slightly lowering pH
- 15-5-15 with calcium and magnesium for increasing pH

Micronutrients (especially iron), magnesium sulfate and calcium nitrate have to be applied on a regular basis every fourth feeding (depending on soil and water analysis). If any nutritional problems seem to appear, a good recommendation is to leach the crop. Call for advice, and get soil analyzed immediately.

If EC is high, don't let media dry out. Apply Subdue at 1oz. per 100 gal. every three to four weeks.

If newly developing leaves become yellow under hot conditions, apply 3 to 4 oz. of Sequestrene 330 every two to three weeks in the early morning hours. Rinse leaves off immediately after treatment to avoid burning.

Feeding schedule

One of the most common problems with ivies is that they're grown too wet in the first part and too dry in the second part of the crop (see also edema control). Ideally, ivies should be grown on the drier side in the beginning and fed after half of the fertilizer solution in the pot is used up by the plant. Ivies prefer evenly moist (not wet!) soil for ideal performance. The feeding schedule has to be adjusted to light, temperature, and plant age on a daily basis.

4. Temperature

The ideal temperature to reach optimum growth rate, habit and flower formation is an average of 68° F (20° C); 68° F (20° C) days and nights result in compact growth. With warm days and cool nights, plants will stretch. With a 75° F (24° C) day and a 61° F (16° C) night, good quality can be achieved, although botrytis might be a problem due to low night temperatures. A good temperature regimen is a 72° F (22° C) day and a 65° F (18° C) night. Start cool morning drop for two hours to 48° F (9° C) beginning before first light for more compact plants, six weeks after planting. A regimen of 66° F (19° C) days and 70° F (21° C) nights results in excellent compact, well-branched plants with no botrytis. With warmer days and cooler nights, the more growth regulators will become necessary, especially on the more vigorous varieties.

5. Growth Regulators

Florel

Apply Florel at 350 ppm to avoid stunting or yellowing; 500 ppm can be too drastic.

Compact varieties need fewer treatments (two to three) than more vigorous varieties, which need three to five.

Plants should be well-established (two weeks after planting), actively growing and moist with no sunshine for up to two hours after application. For maximum branching, rooted cuttings can be sprayed with 200 ppm one week before planting. If necessary, pinch the plants four weeks after planting.

Ideally, Florel should be applied up to six weeks before sale; after that, Cycocel should be applied to keep plants compact, depending on temperature, regimen, variety and overall appearance.

Cycocel

Spray Cycocel at 750 ppm to avoid yellowing (including 1 oz. per 100 gal. Aguagrow). Frequent (weekly) applications at low rates result in a more even crop than application at high rates. Compact varieties might need a few applications; more vigorous varieties need more frequent applications. Cycocel can be drenched at 4,000 ppm after sufficient trials. Cycocel should be sprayed very early in the morning; the soil should be moist.

B-Nine

In warmer climates, Cycocel (750 ppm) can be combined with B-Nine at 1,000 ppm and sprayed one to three times.

Bonzi

Especially in warm climates, Bonzi can be sprayed at 1 to 4 ppm depending on growth habit and temperature, and one to four times after the initial Florel treatments. Bonzi drenches should be done only after sufficient tests. The rate should be 0.25 ppm, with the soil moist and the plants actively growing.

Water stress

Trying to keep ivies compact by growing them dry will slow down the overall development and very often cause severe edema. Water stress isn't recommended as a means of growth control.

6. Branching, Disbudding, Leaf Removal

Branching

Good branching, habit, and flower power is always a result of combining variety characteristics, DIF, light, temperature, feeding schedule, EC in soil, Florel, height control, spacing and pinching. All of those factors combined wisely will result in excellent plants.

Disbudding

If possible, take all of the buds off up to four weeks before sale. This will result in a higher bud count, better branching, and more compact plants at the time of sale.

Leaf removal

If possible, take large, yellow leaves out at the beginning of April to produce a more even, healthier basket. If Florel is applied sufficiently, this might not be necessary.

7. Light

Ivies prefer light levels between 2,500 and 3,500 f.c. Hanging in the air above benches, ivies receive more light, heat and humidity than they like. Good air circulation and an optimum feeding schedule for even moisture in the soil are extremely important conditions to minimize or avoid edema and other cultural problems such as root loss and malnutrition. With drip irrigation, separate compact varieties from more vigorous varieties.

8. Edema Control

Edema is a physiological disorder that mainly affects ivies. Under water-, light-, and pH-stress conditions, cells burst and, depending on the severity, they cause small to large corky spots on the undersides of the leaves. This can resemble spider mite damage.

Edema can be controlled by following these steps:
- Make sure plants don't dry out, especially under hot conditions.
- Don't overwater in January and February when plants are small and light levels are low.
- Try not to use saucerless baskets—remove saucers and snap them on before sale.
- Shade if sudden high light conditions occur.
- Keep pH between 5.1 and 5.6, and make sure iron is sufficiently supplied. (Zonals might show toxicity if pH is lower than 5.8.)
- Feed with calcium nitrate every fourth feeding.
- Heat and vent under humid conditions.
- Provide sufficient air circulation.
- Feed in the early morning.
- With drip irrigation, separate compact from more vigorous varieties.
- Keep relative humidity below 75%

9. Insect Control

Ivies aren't a main target for insects. The most important pest is thrips, which has to be controlled completely before flowering. Azatin + Marvik, Duraguard, Talstar + Orthene, Tame, and Sanmite work well against thrips (also against mites), which should be sprayed again every five days, four times in a row. Cyclamen mites and red spider mites can be a severe problem if they aren't detected early enough. Avid, Pentac, and Sanmite treated every five days for four times can easily eradicate these pests.

10. Fungus Control

Botrytis and pythium are the main fungal problems. Prevent botrytis with good air movement, cleaning, high night temperatures and early watering. Heat and vent under humid conditions. Combination sprays with Daconil at ½ lb. per 100 gal. Other botryticides like Chipco or Ornalin at half rate or Phyton-27 can limit botrytis damage.

Pythium can be reduced by using coarse peat and by not allowing plants to dry out, especially when high salt levels are present. Overwatering can also cause pythium. The best prevention and cure for pythium is a drench with Subdue at 1 oz. per

100 gal. every three to four weeks. There is no problem with stunting or resistance at those rates.

These tips are important for growing perfect ivy geraniums. Having a clear mental picture of exactly how the product should look will help you make the right decision at the right time.

Karl Trellinger is technical support and production director, Fischer USA, Boulder, Colorado. October 1997.

Seed Geraniums the Brodbeck Way

Bruce Brodbeck

In 1985, we at Brodbeck Greenhouse couldn't help noticing all of the volunteer geranium plants growing on the greenhouse floor around our plug seeder. Seeds that bounced from the seeder or plug trays during seeding seemed to be thriving on the ground floor without getting the care we gave to our official geranium crop.

Some of the problems we experienced with our geranium plugs included a tendency for the plant to damp off late in the plug growth stage. When we transplanted these 288 plugs into the finished flats, there was also a problem getting the new transplants to take root. Every year we had more misses in the flats than we liked to see. Consequently, patching the misses was just part of our seed geranium production. Since then we've made a few changes to ensure our production is more efficient and cost effective.

Direct Sow for Flats

We originally sowed into 288 and 512 plug trays. During our initial trials of sowing geraniums directly into cell packs, we concentrated our efforts on achieving exact seed singulation. When seeds cost nearly $0.05 apiece, any doubles quickly pushed aside any economies derived from this method.

We configured our seeder to place one seed in the center of each cell. The flat continued on to be covered with about ¼ in. of coarse vermiculite; then it was watered lightly to keep the seed and covering in place while it traveled to the greenhouse and was set on the ground. Once the flats were placed on the ground, they were given a more thorough watering.

In the past three years, we've changed our top coating to Scott's Metro Mix 360. It flows through our top coater hopper nicely, and we've found that its color change more closely reflects the moisture needs of the flat than vermiculite. With the flats now set down in one of our floor-heated houses, we keep them moist, usually watering them once per day. The heating coils in the floor are kept at about 100° F (38° C), which keeps the soil in the flat in the high 70s. The underground heat helps ensure that the germinating seed is receiving the necessary temperature, which may not be obtained solely with conventional overhead heating methods, as temperatures near the floor often vary from ambient temperatures. (We've experimented with covering these flats with a sheet of thin-mil poly sheeting, but the short germination time combined with low light levels at this time of year didn't seem to make much of a difference for the extra steps it took.)

Keeping the house at a relatively high humidity of at least 90% or more generates quite a bit of condensation on the inside layer of our poly houses. We installed a third inside layer of poly to collect any drips and channel them toward the eaves. Any drips lower the desired soil temperature, and consistent drips will erode the top coating on the flat and wash out the seed.

The seeds germinate in approximately five to six days, and over the next two weeks the likely germination percentage becomes fairly clear (usually fairly close to what's indicated on the package). To account for the expected loss in germination, we sow some "spare parts" in plug trays a few days after the direct sowing, usually in the amount expected not to germinate.

After three to four weeks or when the plant is in its fourth or fifth leaf stage, we move them into our growing house. We use a small portable conveyor to run the flats across. Then we fill in any missing cells or remove any misshapen seedlings with the ones grown in the plug trays. At this stage, savings on labor costs are realized. This process usually requires only half of the labor required for a traditional transplanting line, and with only one or two cells needing replacement per flat, production is nearly three times as fast. Flats are then loaded up and moved out into an area where the temperature is maintained at 75° to 80° F (24° to 27° C) days and 60° to 65° F (16° to 18° C) nights.

Media and Fertilization

The soil we use for geraniums is a 70% peat/30% yellow sand mix. The long growing time plus the rather close plant spacing in our 1203 packs put undesirable pressure on the plant to stretch. This mix helps carry the geranium from germination to shippable size.

For fertilization, we use a peat lite 20-10-20 blend as a constant feed at 50 to 75 ppm throughout the whole growing period. For growth regulation, we start applying cycocel about three weeks after germination at ½ oz. per 1 gal. of water. We make three to four applications approximately ten days to two weeks apart.

The method of application must include a few precautions to achieve desired results. We lower the pressure on the sprayer to about 50% of what's normally used for most applications. Then, being careful not to hit any of the leaves directly on the initial spray burst, we spray in a cascading manner, which lets droplets settle down on leaves.

Direct seeding naturally helps plants begin branching out at an earlier stage, which exposes more leaf surface and makes the Cycocel more effective. Before we refined this method, we were getting overdosed patches in the crop that disguised themselves, looking like viral damage with severe leaf burning.

We don't use any fungicides until late in the growth stage, when we treat with Chipco on a biweekly program along with other plant crops when prolonged periods of cloudy weather persist.

Direct Sow for Pots

We also direct sow into 4-in. round pots in a fifteen-count shuttle tray. For this, we use a peat-lite soil mix containing very little sand. Because plants won't be grown as closely together as the packs, the lighter mix tends to hold more moisture and promote more side growth. We sow one seed per pot and set pots on heated floors to germinate.

After three to four weeks, we move plants off of the heated floors into another greenhouse for growing on. It's at this point that we pull out any pots that didn't germinate. After three more weeks or when leaves begin to cover the soil, we space pots to every other one in the shuttle tray.

We use the same growth regulation and fungicide program as we do for plants. Fertilization starts out the same way as well, and we increase to 100 ppm of 20-10-20 as the pots are separated and grown to maturity.

We use this crop to supplement our cutting geranium crop. Taking longer to come into color than our cuttings, the direct-sown crops usually fill late season sales when cutting geraniums have already been depleted by earlier sales.

As plants mature, we generally see buds and blooms the first week of May. Ideally, our customers prefer them to be shipped with buds and three to four blooms per flat to show the color of the bloom and provide longer shelf life in the garden center. Our bedding plant shipping season begins around the last week of March, and with orders for geraniums, this early shipping gives us an opportunity to use the open floor space to space the remaining crop as it further matures. This extra spacing allows air to flow to the plants and deters the formation of yellow leaves under the top canopy, which can take away from the overall display appearance.

For greenhouse operations that have year-round crop production, the long growing times and extra space required probably wouldn't make this method as feasible. We specialize only in the spring bedding plant season, shipping from the last week of March until the first week of June. Starting these geraniums in the last week of December fills an open time window in our plug production between slower growing crops such as perennials and the larger volume crops such as our annual bedding plant plugs. Geraniums are the first crop to be grown in the greenhouses we use to germinate them in, and other crops aren't scheduled to be grown in those houses until after the geraniums have been moved out, so the growing space requirement really isn't an issue.

With the combination of good growing habits and cost savings, this direct sowing program has worked well for us over the past twelve years. We've tried to expand direct sowing into more crops like impatiens and petunias but haven't yet achieved the same levels of success as we have with geraniums.

Bruce Brodbeck is owner, Brodbeck Greenhouse, Toledo, Ohio. December 1997.

Zonal Geraniums

Gary Vollmer

Over the past few years zonal geraniums have become one of the most controversial crops that bedding plant growers deal with. The unpredictable supply of clean cuttings and mind-boggling array of new varieties have made the growers' choice of varieties and suppliers not only difficult but critical. High-quality cuttings are one key to a successful crop, but the rooting environment and finishing conditions are just as important.

Propagation
Unrooted geranium cuttings can be stuck directly into pots or rooted first in one of a number of propagation medium. Zonals root readily without rooting hormones within a temperature range from 68° to 85° F (20° to 29° C). Some growers use mist or fog or keep the cuttings covered until they root. The base of the cutting must be in an environment that's moist, but has a good supply of oxygen. Wet, low-oxygen conditions at a cutting's base will delay rooting and may cause rot. This problem is especially prevalent in direct sticking situations because the large volume of soil stays wet longer. The other potential problem is botrytis. Botrytis can be prevented by providing good air circulation around the cutting and by using fungicides. Space the cuttings farther apart and place them on a wire bench to increase vertical airflow. There are many fungicides registered for use on geraniums that are effective against botrytis, so use them.

One of the common mistakes many growers make when propagating geraniums is overmisting. Provide only enough water to keep the leaves turgid. Zonals can root well in three to four weeks. Many types of starting materials are available to the grower. These include unrooted cuttings, callused cuttings, rooted cuttings (in many sizes and media), prefinished 4-in. pots, and harvested stock plants in a variety of pot sizes. One of the major concerns when buying geranium cuttings, either rooted or unrooted, is ethylene buildup in the shipping box. Be sure to open the boxes immediately upon receipt.

Finishing

Zonals vary greatly in vigor from variety to variety. Some varieties, like those traditionally produced by Oglevee, are bred for garden performance and provide strong vigorous plants. Others, like those traditionally produced by Fischer, are targeted at pack performance and produce compact plants. Today, these varieties still exist but other varieties are finding a happy medium between pack and garden performance. Due to the wide diversity of plant forms available, zonal geraniums can be produced in a wide assortment of container sizes. Vigorous varieties can fill large containers (8 to 12 in.) with a single plant, while the compact varieties can be finished in small (3 to 4 in.) containers grown pot to pot. This wide range of vigor and form allows zonals to fit many niches for the grower and marketer.

Zonals can be grown within a wide temperature range. Cooler temperatures provide strong, well-toned plants but add crop time. Warmer temperatures reduce time to flower, but produce softer plants. Finishing times for 4-in. zonals can vary from six to twelve weeks. Shorter finish times can be achieved with high light, warm temperatures, and CO_2 injection. To use this forcing technique on vigorous varieties, space and growth regulators are needed.

Growth regulators

Some compact varieties may be produced pot to pot with no growth regulators, but plant quality and consumer value are reduced. My growth regulator of choice for zonals is Cycocel at 1,500 ppm. Some varieties need up to three applications. I've found that one application is adequate on most varieties if the plants are spaced adequately. Florel has been used successfully by stock plant producers to enhance branching and abort flower buds, and some growers are using Florel in finished production as well.

The most serious disease problem of zonals in the finishing house is *Xanthomonas pelargonii* (bacterial blight). This bacterial disease cannot be cured, only eradicated. The symptoms can be masked (but not cured) with Phyton 27, but the plants will eventually die prematurely. The disease can be spread by propagation, splashing water, and contaminated containers and tools. Avoid bringing xanthomonas into your greenhouse by buying only virus-indexed cuttings. Strict sanitation techniques during propagation and planting will insure that the disease, if present, doesn't spread.

Geraniums will continue to be a mainstay of bedding plant growers. Seed geraniums are less expensive to produce, while zonals provide bigger doubled flowers and a broader range of forms and colors. Differentiating seed and zonal geraniums to the consumer will help insure a strong niche for both types.

Gary Vollmer is production manager for Harry Smith Gardens in Bellingham, Washington. March 1995.

Control Edema on Your Ivies

Bob Michael

Edema is a physiological problem occurring mainly on ivy geraniums. It is a bursting of the cell walls on leaf undersides. It has a corky appearance that sometimes resembles spider mite damage. Edema isn't a disease, bacterium, or a virus; it isn't transmitted from one plant to the next. It's caused by improper management of the greenhouse environment and overwatering of crops.

Most growers are cropping their ivies in the air above benches. This is a very humid (air) environment. Also, many are growing their ivies in a saucerless hanging basket. These baskets hold a good bit of water after each irrigation, giving the ivies an overabundance of water. Baskets that are hung in the air are on an automatic watering system. When a few plants start to dry out, the entire crop is watered, whether the rest of the crop needs to be irrigated or not. Air circulates poorly where the baskets are hung, reducing the transpiration rate.

Finally, ivies are grown in a low light time of the year during January and February, again reducing the transpiration rate. These two months are also the coldest, which reduces the air temperature in the greenhouse. This is the perfect scenario for edema to occur. Moisture is so heavy in the media and in the air that plants can't transpire. In a nutshell, ivy geraniums are drowning! The natural defense mechanism to avoid drowning is for cell walls to rupture so moisture can be purged from plants.

What Can Growers Do to Avoid Edema?

- Properly manage the greenhouse environment. Reduce air humidity by venting your greenhouse first thing in the morning, even if that means turning up the heat (a decondensation program).
- Make sure there is adequate airflow, whether it be through heat jet fans or horizontal airflow. Plants need air movement twenty-four hours a day.
- Don't use a saucerless hanging basket. Go back to using baskets where the saucer snaps to the outside of the basket, but don't put the saucer on until the point of sale. This will ensure maximum drainage of each basket.

- If you're using an automatic watering system, place varieties with similar vigor on each line or section, again to eliminate overwatering.
- Manage your pH properly. Make sure your ivy pH is 5.5. When you keep the pH at 5.5, you can feed once every three feedings with calcium and potassium nitrate. Calcium will thicken up your cell walls, making ivies more resistant to edema.

Bob Michael is technical services manager, Oglevee Ltd., Connellsville, Pennsylvania. March 1996.

Gerbera

Gerbera

Tom Linwick

Gerbera has come a long way since it was introduced as a pot plant more than a decade ago. Use of tissue-cultured gerbera has been on the decline recently due to some of the tissue culture labs discontinuing production and the increased quality of seed-produced gerbera.

Gerbera can be used as both a pot plant and a bedding plant. They work well in mixed containers, as flowers are somewhat unusual and very attractive. Most growers produce gerbera in 4- to 4½-in. pots so they can produce more plants per square foot of bench space, but gerbera is also very suitable for 5- and 6-in. pots.

Propagation

Gerbera can be sown directly into a variety of larger plug trays from an 84 tray up to a 200 tray. Some growers prefer to sow into a smaller plug tray and transplant up to the larger tray after about three weeks. By transplanting, you can grade seedlings according to their size, but this adds some time to the crop. You can also broadcast seeds into open seed flats; however, this requires more work and delays plant development.

Seed flats

Seedling media should be a slightly fertilized peat with 20% perlite added for good aeration. Soil pH should be 5.5 to 5.8 with a 1.2 to 1.5 EC. Water all plug trays before sowing to ensure adequate moisture. After sowing, cover seeds with a thin layer of coarse vermiculite.

Temperature and humidity

Place seed trays in an airfog system at 75° F (24° C) for twenty-four hours, then cover them with plastic to keep humidity levels as close to 100% as possible. Put seed trays into a germination chamber at 73° to 75° F (23° to 24° C) for four to five days.

When most seeds have germinated, remove the plastic and move seed flats into the greenhouse. The greenhouse temperature should be 68° to 72° F (20° to 22° C). (Higher temperatures are preferable.) At this point, keep seedlings away from direct sunlight by using a light shade. Greenhouse humidity should be at 85% to help the last seeds finish germinating. Seeds require four to six days to germinate, depending on germination temperatures.

A variation of this method is to germinate seed trays in the greenhouse without a germination chamber. You should still maintain the proper temperature and high humidity during germination. Also, some growers prefer not to cover seeds with vermiculite. This is acceptable if you maintain proper humidity levels.

When all seeds are germinated and well-established, keep humidity between 70 to 75% to prevent fungus attack. Higher humidity levels will also cause stretching of young plants. Good air circulation and ventilation during the day is essential for better quality plants.

Lighting

When seed flats are in the germination chamber, neon lighting is beneficial. Alternate light with one hour on and one hour off for twelve hours of each day. The light intensity should be 20 to 30 watts per square meter.

When seed flats or plug trays are in the greenhouse, use supplemental lighting if day length is less than fourteen hours. The amount of light required is 60 to 80 watts per square meter, or 300 to 500 f.c. if measured with a conventional light meter. Under higher light conditions, gerbera will form more flower buds.

Fertilization and watering

When seedlings are approximately ten days old, you can fertilize them with a half strength feeding of 15-5-15 or 15-15-18 at 50 to 75 ppm. After about fourteen days it's beneficial to let seedlings run somewhat dry to prevent plants from becoming stunted with thick leaves and darker foliage.

Gradually increase amount of fertilizer until the rate is up to 100 to 150 ppm with each feeding. Watch soil pH so it stays below 6.0. If soil pH is too high, leaves may become discolored or chlorotic due to iron and manganese deficiency. Low temperatures can also result in discolored leaves.

Growing On

Gerbera is a relatively long crop, requiring sixteen to twenty weeks to finish, depending on the time of year and the climate for production. You can produce it year-round, but finishing in winter is somewhat difficult, especially in the northern latitudes, where supplemental lighting is necessary.

Transplanting/potting

When plants are six to seven weeks old, they should be at the four to five true-leaf stage and ready for potting into their final containers. An additional week of production time is required to reach the four to five true-leaf stage during winter. When receiving gerbera plugs or liners, place them in the greenhouse and allow them to acclimate for one to two days. Pot plants as soon as possible to prevent root systems from becoming overgrown or root bound. This can add time to finishing the crop and result in smaller plants at flowering.

When transplanting gerbera, be careful not to cover the plant crown, which branches at the soil line. Plant slightly higher than the soil line in pots.

Potting soil

The soil should be loose and well drained with 20% perlite added for proper aeration. Other soil amendments such as vermiculite, calcine clay or coarse sand are also beneficial. Gerbera root systems won't develop properly without good aeration. A slightly fertilized peat will accelerate root development. The pH should be 5.5 to 5.8.

Temperature

A 68° F (20° C) night temperature is preferable until plants are well-established in their final containers. When first transplanted, gerbera requires a minimum night temperature of 65° F (18° C) to help establish a good root system. Day temperature should be around 75° F (24° C). Extremely high or low temperatures will delay flowering. With a slightly higher temperature, plants will finish more quickly and evenly.

You can also use negative DIF with a 65° F (18° C) day temperature and a 72° F (24° C) night temperature. Another method is to lower the greenhouse temperature for three to four hours after sunset. This helps reduce any possible stretching. When flower buds are visible you can lower the night temperature slightly, but this will delay flowering. Maintain a minimum night temperature of 60° F (16° C) to ensure proper development. Extremely high or low temperatures will delay flowering.

Humidity

Keep humidity at 70° to 75° F (21° to 24° C) to help prevent fungus attack. Good air circulation and ventilation during the day are essential for good plant quality. High humidity will cause plant stretching.

Lighting

Gerbera requires high light levels for top quality plants. For this reason, most are produced during spring and summer, with the heaviest time of production in spring. They produce more flower buds with high light levels than under low light conditions.

If the day length is less than thirteen to fourteen hours, supplemental lighting at 40 watts per square meter is beneficial. Supplemental lighting will increase plant quality, especially during winter.

Fertilization

When plants are young, use lower levels of feeding, then increase as plants mature. In the early stages use a complete fertilizer at 100 to 150 ppm, then increase to 150 to 200 ppm near the crop finish. Some peat-lite mixes, such as a 15-5-15 or a 15-15-18, work very well. Avoid high ammonia-type fertilizers, as they'll produce more stretched foliage. The EC should be 1.5; the pH should be 5.5.

Watering

Gerbera should receive thorough waterings, then allowed to dry somewhat to prevent stretching. By allowing plants to dry out between waterings, you can avoid the need for growth regulators. Also, be sure to water early in the day to ensure that foliage is dry at the end of the day.

Spacing

After potting, plants can remain pot to pot for approximately four weeks. At that point you need to space plants to allow light to reach the plant crown. It's important that light reach plant centers for satisfactory flower bud development. If plants don't receive enough space during production, the result will be delayed flowering, fewer flowers, and larger, long leaves.

Shading

A light shade may be necessary during summer to keep temperatures at more acceptable levels. Be careful not to shade too heavily or too much, as gerbera is a high light plant and too much shading may cause stretching.

Growth regulators

You can use B-Nine after plants are at the four- to five-leaf stage at 1,000 to 1,500 ppm. When plants are established in their final containers about ten to fourteen days after potting, you can spray with B-Nine at 2,500 ppm. Growth regulators aren't necessary if you can grow plants on the dry side. They're useful if you're finishing smaller pots (4 in.), and you can't run plants dry. Be careful not to use B-Nine too close to flowering or in the last three to four weeks of production because it can affect flower stem length.

Pest and diseases

Watch for leafminers, thrips, whiteflies, cyclamen mites, and aphids. Many new chemicals for pest control are available, and some have longer-lasting control. Some diseases to watch for are alternaria, phytophthora, and mildew. Botrytis can also be a problem on open flowers.

As you become better at watching the details involved in growing gerbera, you can reduce crop times and increase plant quality. The demand for quality gerbera hasn't been met by growers in many areas. Gerbera is a bit more challenging to grow than some pot plants, but the return on a quality plant can be excellent.

Tom Linwick is technical representative, Daehnfeldt Inc., Duvall, Washington. January 1997.

Horticultural Oil for Powdery Mildew on Gerbera

David L. Clement, Rondalyn M. Reeser, and Stanton Gill

As part of an ongoing greenhouse TPM/IPM program at the University of Maryland, we conducted trials in 1996–97 to evaluate the effectiveness of several horticultural oils for preventing powdery mildew, *Erisiphe cichoracearum*, on gerbera daisy.

We evaluated three horticultural oils on gerbera daisy cultivars Festival Mix and Lemon Yellow between August 1996 and March 1997. The oils tested were Stylet Oil (JMS Flower Farms Inc., Vero Beach, Florida), SunSpray Ultra-Fine Oil (Sun Co. Inc., Marcus Hook, Pennsylvania), and Triact 90 EC (Thermo Trilogy Corp., Columbia, Maryland).

These products were applied to runoff on four-week-old transplants at rates of 0.5%, 1%, and 2% (v/v) over a six- to eight-week period. The standard fungicide control was Cleary's 3336 applied at the label rate.

Sprays were applied at one- or two-week intervals. Plants were rated weekly during the trials for disease incidence. Temperatures during the spray trials ranged from a low of 60° F (16° C) at night to 108° F (42° C) during the day. Relative humidity values ranged from a low of 26% at night to 74% during the day.

Results

Horticultural oil treatments affected flower initiation, foliage color, plant height, and vigor. Plants treated with Cleary's 3336 appeared normal, had little visible residue, and showed no disease when sprayed at weekly intervals throughout the trials. Some plants sprayed with Cleary's 3336 at two-week intervals did develop powdery mildew. Plants treated with Stylet Oil at all spray rates and intervals had consistently smaller leaves, shorter height, and lighter green foliage compared with the other oil treatments. Plants treated with Triact 90 EC at all spray rates and intervals usually had the largest leaves, greatest height, and darkest foliage. Plants treated with SunSpray Ultra-Fine and Stylet Oil consistently flowered earlier than those in other treatments.

Regardless of the season, applications of SunSpray, Triact 90 EC, or Stylet Oil at the 2% rate applied either weekly or every two weeks, caused a reduction in leaf size, reduced plant height, and prevented flower initiation. Additional treatment effects for all oils at the 2% rate included a shiny residue buildup on the leaf surface, lighter green foliage, and earlier leaf senescence.

We speculate that the 2% rate of oil caused tissue damage, reduced respiration rates, slowed the growth rate and additional plant stresses. Oil treatments at the 1 or

2% rate at temperatures consistently above 90° F (32° C) caused more problems with plant size, color, and flowering.

The highest rate of oil that could be used without affecting plant size or color was 1% every week during temperatures less than 90° F (32° C). Applications of SunSpray, Triact 90 EC, or Stylet Oil at the 1% rate every two weeks gave few problems with phytotoxicity symptoms and still gave adequate disease control.

All three oils applied at the 0.5% rate at two-week intervals gave the fewest phytotoxicity symptoms, but disease control was inadequate in some cases. Triact 90 EC applied at the 0.5% rate at two-week intervals gave good control of powdery mildew. The 0.5% rate of SunSpray Ultra-Fine Oil or Stylet Oil at two-week intervals didn't consistently control powdery mildew.

Based on our trials, the grower incorporated the use of 1% horticultural oil rates of SunSpray or Triact 90 EC at one- or two-week spray intervals depending on the severity of disease throughout the remaining 1997 growing season with good success.

David L. Clement, Rondalyn M. Reeser, and Stanton Gill, University of Maryland, Ellicott City. April 1998.

Godetia

Godetia: Grower's Guide to Perfect Cut Flowers and Pot Plants

Bob Anderson

Once a unique and uncommon garden plant, godetia (satin flower) has become a dependable cut flower for greenhouse and field production and a reliable choice for pot plant production. Whether you're producing this now-popular crop for cut flower or pot plant markets, our cultural guide will help you succeed every time.

Godetia is infested by aphids, thrips, whiteflies, and spider mites when these insects are a problem on other plants in greenhouse or field production. Root- and stem-rotting diseases such as fusarium and pythium can be a problem when these plants are grown in poorly drained soils or when they're overwatered. Gray mold (botrytis) can cause leaf and stem damage on tightly packed, overfertilized plants in the greenhouse.

Plug Production

The first four weeks of godetia production for cuts, or six weeks for pots, are best completed as plugs. Seed germination is usually uniform and at high percentages. Seed is often single sown into 288 or similar-sized plug trays for cut flower production or larger cells for pot production.

Stage 1, Days 1 to 10

Sow seed into plug media with little or no starter charge. Lightly cover seed with media or vermiculite; maintain uniform soil moisture with intermittent mist, and maintain media temperatures of 70° F (21° C) for seven to ten days, until germination is complete.

Stage 2, Days 11 to 21

Move seedlings to a cool, bright, well-ventilated greenhouse; optimum temperatures are 55° to 60° F (13° to 16° C). Fertilizer applications directly affect the number of lateral branches. Use almost no fertilizer (beyond a starter charge in the plug media) if plants will be grown as single-stem cut flowers. Fertilize pot crops and multiple-stem cut flower crops one or two times during plug production to get four to eight lateral branches per plant. Apply only 50 to 100 ppm fertilizer at one time. Don't use constant liquid fertilizer applications, or the crop will be too soft. Use a well-balanced fertilizer with calcium, or consider using a balanced organic fertilizer. Supplemental HID lighting for six to twenty-four hours each day will greatly enhance plug growth in low light areas of the country.

Stage 3, Days 21 to 28

Maintain cool temperatures. Negative DIF or weekly applications of B-Nine (2,500 ppm) will help control plant height for pot production. Cool temperatures and wise use of water and fertilizer will make the best plugs.

Stage 4, Day 28

Plugs are ready to transplant or ship. Transplant plugs as soon as possible. Obtain plugs for cut godetia crops for Thanksgiving or Christmas from a cool climate, probably California, because seedlings don't tolerate warm temperatures in the plug stage.

Field Cut Flower Production

Most field-grown godetias are produced on the West Coast because uniform cool temperatures are common during production. Field production is possible in the upper Midwest, New England, and in the mountains with some experience. California growers harvest sequential crops from May into July from fall, winter, and spring sowings. Crop time and stem length decrease as average temperature, light intensity, and day length increase from spring into summer.

Plants can be planted in rows 5 ft. apart with in-row spacing of 2½ to 3 ft. Large individual plants with twenty to fifty branches are produced, and most cut stem lengths will be 14 to 20 in.

Cut stems will be 20 to 36 in. long if plants are planted in double rows with a spacing of 9 to 12 in. between plants. Transplants are pinched at transplanting to remove the primary stem and leave four to eight lateral branches. These rows must be supported in the field.

Outdoor production requires more fertilizer than greenhouse production but still much less than most crops. Use low-fertility soils with good drainage. Keep soluble salts below 0.6 mmhos. Organic fertilizers have been successful in field production in California. Be careful not to damage plugs at transplanting, and don't plant too deeply.

Greenhouse Cut Flower Production

Godetia is a spectacular greenhouse cut flower crop with long stems and large flowers. However, the crop must be grown with almost no fertilizer through the twelve to sixteen weeks the crop is in the greenhouse. Godetia can be grown in ground beds, but it must be nearly free of fertilizer. Godetia planted in beds following other crops or where bedding plants or pot crops were grown will probably fail due to the soil's relatively high fertilizer residue. Overfertilized plants are soft; their stems are crooked; there are many lateral branches along the stem; and plants are easily knocked down with watering and difficult to support.

Even so, godetia is a great candidate for cut flower production in pots. Cut stems 24 to 40 in. long can be grown from single-stem plants grown as one plant per 4-in. pot or two to four plants per 6-in. pot, with a plant density of eight to ten plants per square foot. Try not to set these pots on ground beds, where they'll root into the soil below and deteriorate because of soil fertilizer residue.

If you use ground beds and soils are low in fertility, pinched plants with four to eight lateral branches could be planted at a density of one to two plants per square foot. Pinched plants will require two to three weeks more production time than single-stem plants. Plants will require one to two layers of support. Younger plants, eight weeks old to when buds are 1 in. long, are softer and easy to knock down with careless overhead watering. Always water soil rather than foliage. It's best to keep plants on the dry side, however; wilted plants will have permanently crooked stems when they recover.

Supplemental lighting can speed godetia winter flowering because it's a faculta-tive long-day plant. Flowering in the Grace series of cut flower godetias can be speeded four to six weeks using incandescent (mum) lighting or HID (200 to 400 f.c.) supplemental lighting from 6 P.M. to midnight each day during production. Using HID supplemental lighting for twenty-four hours per day for four to six weeks after germination will also reduce production time significantly. Best winter greenhouse production will involve plugs grown under four to six weeks of HID supplemental lighting for twenty-four hours per day.

Scheduling depends on greenhouse temperature, supplemental lighting treat-ments, time of year (total light accumulated by the crop), and variety. In Kentucky, four-week-old plugs transplanted in late September flowered in mid-December (eleven to twelve weeks, for Christmas); plugs transplanted in mid-October flowered in early February (fifteen to sixteen weeks, for Valentine's Day); transplanted in early November, they flowered in mid-March (seventeen to eighteen weeks). These results all came when supplemental incandescent lighting was used for the whole crop in a greenhouse with 50° F (10° C) night and 60° F (16° C) day temperatures.

Godetias are excellent cut flowers. Grown properly, stems are strong and straight. Each stem will have four to fifteen flower buds, depending on the plant's overall vigor. Cut stems have a vase life of fourteen to eighteen days, and all flower buds open to normal size and color when flowers receive proper care in a flower arrange-ment. Research has shown that floral preservatives with sucrose will damage leaves and reduce vase life. Cut stems in tap water performed equally well or better than in preservatives in vase life trials.

Pot Plant Production

Godetias in the Satin series are dwarfs suitable for pots. Overall plant size is reduced significantly, and the flower size is slightly reduced, compared to cut flower types. Satin godetia is well suited for 4- to 6-in. pots and is grown as one plant per pot.

For pot production, it's easier to transplant a larger plug (from a 1- to 1½-in. cell) that's approximately six weeks old. A late December or early January sow date should result in plants for Mother's Day in most of the northeastern and midwestern states. Keep plants in a cool greenhouse at 50° to 55° F (10° to 13° C) night and 60° to 65° F (16° to 18° C) day temperatures, preferably with negative DIF, after transplanting. Although the growth retardants B-Nine, Cycocel, Bonzi, and Sumagic, are effective, they're not necessary when godetia is grown cool, dry, and with very low nutrition.

Watering and fertilizing practices are critical to pot godetia production. Careless overhead watering will weaken the plants and open the plant canopy. Consider sub-irrigation or drip irrigation. Use very little fertilizer on plants after transplanting. A single application of a ¼ tsp. of Osmocote 14-14-14 at transplanting was sufficient for 4-in. pots of Satin godetias in our trials. Consider applications of 50 to 100 ppm from a balanced liquid fertilizer every fifteen to thirty days or a balanced organic fertilizer as you evaluate production practices for pot godetia. To prevent stem stretching be sure individual plants always have sufficient space as they develop.

In spring California-grown dwarf godetias are shipped east for combination pots and other uses. Dwarf godetias are as tolerant of cold temperatures as pansies, snapdragons, and dianthus. The unique flower color patterns of godetia make it an excellent addition to early spring bedding plant sales.

Godetia may be flowered as a pot plant for Mother's Day and early summer using the Satin series. Following is a production schedule from Satin's breeder, Sakata Seed of Morgan Hill, California, and Yokohama, Japan.

Stage 1, Days 1 to 10

Single-sow into a large plug cell (128 is ideal), using a plug media with a low starter charge. The larger plug cell allows more natural light around the plant, helping to reduce stretch and increase basal branching. Cover the seed lightly with either media or vermiculite, and maintain a soil temperature of 70° F (21° C) with even soil moisture.

Stage 2, Days 11 to 21

As soon as one sees green in the plug tray (around ten days), immediately move the tray to a cool, bright, well-ventilated greenhouse. Supplemental lighting can benefit the plug and ensure its healthy development. Optimum temperatures are between 55° and 60° F (13° and 16° C). The use of a negative DIF of 55° F (13° C) day, and 65° F (18° C) night or a two-hour temperature drop of 5° to 10° F at daybreak, followed by moderate day temperatures of 60° to 65° F (16° to 18° C), is ideal for godetia production. If the plug media has no starter charge, feed plugs lightly with 50 to 100 ppm nitrogen, preferably from a well-balanced, calcium nitrate–based fertilizer. Negative DIF combined with low media fertility will keep plugs compact and toned. High day temperatures, high media fertility, and low light will weaken plugs, resulting in soft, leggy growth.

Stage 3, Days 22 to 40

Maintain cool temperatures, and use a negative DIF, if possible. Weekly sprays of B-Nine (5,000 ppm) will help control plant height, but temperature manipulation has proven to be the most effective tool. Feed plugs lightly, using 50- to 100-ppm nitrogen every ten to fourteen days from a well-balanced, calcium nitrate-based fertilizer. Using high ammonium- and urea-based feeds is *strongly* discouraged.

Stage 4, Day 40

Plug trays are now ready for transplanting or shipping. They can be held in a well-lighted area at 40° F (4° C) to tone before shipment or until space is available for planting.

Four-in. pot to eight weeks

Godetia plants are sensitive. Dislodge them from the plug tray by pushing up from the bottom. Avoid pulling plants out of the tray by hand, which may damage the stem. Avoid planting plugs below the soil line to guard against stem rot and ensure a healthy transition.

Maintain cool growing conditions of 55° to 60° F (13° to 16° C), preferably employing negative DIF.

Ideally, use drip tubes or subirrigation, especially when in flower. Overhead watering with strong water pressure will weaken plants and open up the plant canopy.

You have two fertilization choices: liquid and dry feed. With liquid feed, fertilize every fourteen days with 100 ppm nitrogen from a well-balanced fertilizer, like 15-5-15 or 20-10-20. With dry feed, top-dress each pot with a ¼ tsp. of 14-14-14 Osmocote with *no additional fertilizer required*. (Note: Plants that appear undernourished during the middle of the production cycle will have stronger stems. Plants can be greened up in the last two weeks as flowers begin to open.)

Maintain a media pH between 5.5 and 6.5. Many growers rely on the acidifying properties of liquid fertilizers to control media pH. Because little fertilizer is being used, especially with the Osmocote-only method, acidification of irrigation water to reduce alkalinity to 100 to 120 HCO_3 may be necessary to maintain optimum media pH.

Godetia Satin is responsive to daylength, and extending the daylength to midnight (six hours) using ordinary mum lighting (10 f.c. from incandescent bulbs placed on 6-ft. centers) will hasten development.

Godetia plants need to be spaced to ensure good plant quality. Acclimated plants of godetia Satin can tolerate light frosts, permitting spacing outdoors in late April or early May, when greenhouse bench space is at a premium.

For garden performance, godetia Satin does best under mild weather conditions. In areas where summer temperatures regularly exceed 80° F (27° C), plants will benefit and perform better if given shade during the hot afternoon.

Bob Anderson is extension floriculture specialist, University of Kentucky, Lexington, Kentucky. November 1996.

Herbs

Sage Advice for Herb Production

At Cascade Cuts, Bellingham, Washington, owner and grower Alison Troutman grows about 28 of the 750 sages available. Of those, perhaps 10 are the hottest culinary choices, she says. The ornamental and more medicinal ones are gorgeous plants that tend to not bloom till late summer. As ornamentals, they have their most value as potted tub or conservatory plants for fall to winter sales. In a seminar at GrowerExpo '98 held near Chicago in January 1998, Troutman revealed her tips for success with sage.

Propagation

Garden sage is the only one they produce from seed. Bear Garden, a new cultivar, has large, juicy leaves. They sow in 200- or 125-count trays, making three passes for three seedlings in each cell. That makes a nice short, fat, bushy plant. Crop time is about seven weeks during the cooler months.

For vegetative propagation, you need to take sage cuttings while the growth isn't too soft, says Troutman. Once you have the right kind of growth, almost anything goes. Once you stick it, you can even bury the two lower leaves. Leave as many leaves on as possible, because you've got to fill the pot with leaves before you sell it. Most varieties can be finished in five to six weeks during active growing months.

Most of the garden sage types are fairly hardy. The colored sages (red, golden, and tricolor) are all derivations on the theme of garden sage. If you want to do a few fragrant varieties, do pineapple sage or honeydew melon sage. They're fruity, flavorful varieties that can be used in salads.

Pests and Disease

Sages can be susceptible to aphids and mites, especially during the tender growth phase, Troutman says. Thrips can also occasionally be a problem. Treat with Botanigard, Garlic Barrier, or Hot Pepper Wax during cool, early growing months and later with Empede or SunSpray Oil.

Mites are often the only problem Cascade sees all year, says Troutman, so they plan ahead with an early release of the beneficial mite *Phytoseiulus persimilus*. A tip: Keep humidity up around sages, and mites will be discouraged.

Powdery mildew can be a problem—it generally strikes in the early fall months. If the plants are stiff and hardened, if they've been outside, or if you know that the foliage is really tough, you can use a baking soda spray. But be very cautious: If the foliage is too soft, it will burn. Do a test sample first at low concentration. If you don't feel confident of the quality of the foliage, use a SunSpray Oil-Azatin mix.

Culture Notes, April 1998.

Best Basil Tips

The best basils are those grown organically, Cascade Cuts' Alison Troutman recently advised at GrowerExpo '98 in Rosemont, Illinois. For success with basil, innoculate soil with beneficial microbes/fungus with organic matter, as basil is constantly hungry, she says.

On bright days during spring, foliar feed with fish and kelp to give plants a boost. Don't give in to demands to supply basil too early—you'll only get complaints of basil damping off. Instead, grow it when it prefers, in the warmer months. Try varieties such as African Blue or African Cinnamon during cooler months, as they're more perennial in habit.

At Cascade Cuts, they sow every other week throughout spring and most of the summer. They grow only two varieties from cuttings: African Blue and African Cinnamon. Growing time is four to six weeks, depending on the time of year.

As for pests, fusarium, pythium, and rhizoctonia are the diseases to consider. Aphids and thrips can also be a problem. Aphids aren't easy to treat on a quick-turn crop such as basil. Troutman hasn't found releasing beneficials on basil to be economical because of its quick turn,  but you can use them indirectly. For example, luckily, on a crop of scented geraniums nearby, Troutman maintains a population of *Aphidoletes aphidimyza*. If aphids attack the basil, the *Aphidoletes* adults fly to the basil to feed on the aphids.

If you don't have this opportune setup, spray the whole crop with Garlic Barrier as a repellent preventative. For more severe infestations, be prepared to use horticultural oil or soap, but beware of phytotoxicity on basil's tender leaves. Use oil only on sunny days, and use soap only on overcast days.

As temperatures increase with the approach of spring, thrips may also be a problem, and their damage can be devastating on basil. Be sure to keep neighboring crops and overhead baskets pest-free during this high turnover period. Monitor crops twice per week. Healthy populations of *Cucumeris* on surrounding crops will minimize the risk of thrips outbreaks.

Some basil varieties Troutman uses include:

Ocimum basilicum
- Cinnamon, a slow starter that has a reddish stem and is spicy and flavorful,
- Genovese, a popular large-leafed type that's great for pesto,
- Licorice, which can be a bit erratic and has poor germination,
- Mrs. Burn's Lemon, an improved variety of lemon basil that's easier to grow,
- Red Rubin, which is an improvement over Dark Opal but is still too slow,
- Siam Queen, a new Thai basil and All-America Selections winner that has great flavor but is still susceptible to fusarium.

Others
- *O. kilimanshcarium* x *basilicum pururascens* African Blue, which makes an attractive shrub
- *Ocimum* hybrid African Cinnamon, a new hybrid that isn't officially named.

Culture Notes, March 1998.

Lavender Deciphered

Miriam Levy

The genus *Lavandula* contains numerous species, many of which are increasing in popularity for use as annuals and perennials. There are many choices, but what are the differences? How do you choose which ones to produce? Seed or vegetatively produced, gray or green foliage, deep purple, light blue, pink or white flowers, dwarf, tall, winter hardy or semi-hardy, Spanish, English, or French—let's decipher!

All lavender species produce seed, but the most common production method is to produce cuttings vegetatively. With the exception of Lavender Lady, most seed-produced types are not true hybrids, and finished plants vary considerably. Starting with vegetatively produced plants ensures a product that is consistent in leaf color, growth habit, and bloom color.

The most common lavender is *Lavandula angustifolia,* or English lavender. This is the species from which all others were bred, the true garden fragrant type and the lavender used in perfumes and potpourri. Munstead and Hidcote are named varieties of angustifolia. All have gray foliage with smooth leaf margins. The major difference between Munstead and Hidcote is growth habit. The most popular Munstead type is Lavender Lady, a seed-produced variety that blooms the first year. It should be treated as an annual. Munstead as a rule will be taller in the garden than Hidcote, 12 to 24 in. versus 12 in. for Hidcote. Flower color is also different. The Hidcote type available is a Jean Davis variety with light pink or purple flowers; Munstead has deep purple blooms. True English lavender can reach 3 ft. in the garden and has blue flowers. Bloom time is summer and fall for Hidcote and Munstead, spring and fall for angustifolia. All require full sun with well-drained soil.

French lavender, *Lavandula dentata*, has a different leaf shape than English lavender. French has green or gray foliage with a deep serration on the leaf margin. It's more sensitive to cold temperatures and thus less winter hardy. Its primary use is as topiary and indoor pots. *Lavandula dentata* will grow to 3 ft. in the garden and has blue flowers with a bloom period of spring and fall.

Lavender variety characteristics

Variety	Leaf color	Flower color	Height	Bloom time
L. angustifolia	Gray	Blue	3 ft.	Summer/fall
Hidcote Blue	Gray	Deep blue	12 in.	Summer/fall
Hidcote Pink	Gray	Pink	12 in.	Summer/fall
Lavender Lady	Gray	Deep blue	12 to 24 in.	Late spring
Jean Davis	Gray	Light pink	12 in.	Summer
Munstead	Gray	Purple	2 ft.	Summer/fall
L. dentata	Green	Blue	3 ft.	Spring
Gray	Gray	Blue	3 ft.	Spring
L. stoechas	Gray	Purple	3 ft.	Summer
White	Gray	White	2 ft.	Summer
Provence	Gray	Deep purple	2 ft.	Summer/fall
L. pinnata	Green/gray	Dark blue	2 ft.	Spring

Spanish lavender, *Lavandula stoechas*, has the showiest flowers of all the lavenders and is the most vigorous grower. Two varieties of stoechas are becoming very popular, Quasti and White. Both have gray foliage, and Quasti has a very showy purple flower that will reach 3 ft. in the garden. White will only attain 2 ft. in the garden and, as its name implies, has white flowers.

Lavandula pinnata is very tender and is used in the landscape only in very mild southern climates. It is not winter hardy in the U.S., but when treated as an annual it will bloom from spring through the first frost. Pinnata has gray green foliage and deep purple blooms and will reach 2 ft. in the garden.

Intermediate hybrids add to the confusion. These are crosses between hybrids of angustifolia and latifolia. Provence is similar to angustifolia in fragrance but is a larger, taller plant with a light-colored flower. Because of its vigorous growth, Provence is used in the garden primarily as a hedge. Goodwin Creek Gray is a cross of dentata and woolly lavender. It looks like a dentata, with gray foliage and deep blue flowers. In warm climates Goodwin Creek will bloom year-round. It's a good southern variety.

The best time to propagate lavender is late summer and early fall. Cuttings benefit from bottom heat of 65° to 75° F (18° to 24° C). As with most silver-leafed plants, too much mist will cause leaves to rot.

Lavender requires full sun and a well-drained, porous soil during production and in the garden. They are light feeders; a 20-10-20 fertilizer every few waterings is sufficient. In southeastern and Mid-Atlantic states high relative humidity leads to their decline. They don't perform well in warm, humid areas and should be treated as an annual in these locations. Winter-hardy lavenders don't like to be mulched. This keeps moisture levels too high, and they'll rot. In northern states, overwintering success is determined by the size of the root system before the plants become dormant. Most varieties benefit from a spring planting so they can establish themselves before winter temperatures drop.

Miriam Levy is Ball Seed sales representative for Southern California, Encinitas, California. May 1996.

Herb Culture Tips: Lavender

Producing cuttings is the most difficult aspect of managing lavenders, said Alison Troutman of Cascade Cuts, Bellingham, Washington, in a seminar at GrowerExpo '98. She offered tips for conquering that and other challenges of lavender in her herb production discussion.

Stock Plants

These plants come up in the spring, and once they start to elongate and grow vegetative growth, the window for taking cuttings is narrow. After the buds set, the cutting take drops about 40%, says Troutman. While the buds are still at a young stage, you can nip out the buds and continue to take cuttings. But the desire for them to bloom is so strong that your timing has to be impeccable. You can also take summer and fall cuttings—there are never enough! Water them in well, and back off on the mist as soon as possible, Troutman says. Don't overwater.

Seed

The two varieties—Compacta and English—that Cascade Cuts grows from seed are sown into 392s. They generally make three to four passes through the seeder on a tray, and shoot for three good seedlings from a plug, even in a 392. The seeds germinate in just three to five days, are put into 70° F (21° C) and grown on for a couple weeks, then, before they stretch, seedlings are put into a 60° F (16° C) house and continue to root out, says Troutman. After four to five weeks they are transplanted.

When seedlings are ready to transplant, they're moved directly to a cold frame. They stay in 2½-in. pots, grow vigorously because it's the first week of September, grow out, flush out, and cool down naturally as the nights begin to cool. They'll go into a dormant phase and stay pot-to-pot and leaf-to-leaf in 2½s all winter, Troutman says.

Growth

Flexibility and foresight are two key words in lavender production, she says, as it's such a slow crop: "You've got twelve weeks to get a nice pot, so take your time."

Difficulties

Troutman sees few insect problems with lavender. Diseases, however, particularly pythium, fusarium, and rhizoctonia, are a concern. Protect against these diseases, she advises. When they stick cuttings, they use Soilguard 12-G mixed into plug trays.

Culture Notes, May 1998.

Hypericum

Growing Cut Hypericum

Jeff McGrew

Hypericum species have been part of the landscaping and perennial garden for years. Most hypericums flower under long-day conditions and are considered long-day plants. Recently, *Hypericum androsaemum* and *H. inodorum* species and hypericum hybrids have been used by professional cut flower growers to produce a unique and colorful filler for flower arrangements.

Varieties

The most common hybrids are Autumn Blaze, which has clusters of reddish brown berries, and Excellent Flair, which is quicker to flower than Autumn Blaze. Excellent Flair has fewer berries that are brown to red in color, however, and is also more susceptible to rust than Autumn Blaze. Both perennial varieties are hardy to about Zone 5. Extreme heat and humidity are detrimental to plant survival and quality.

Propagation

Propagation of these varieties is vegetative, normally by rooted cuttings. It's important to build the young rooted plant properly so its growth habit will yield maximum stem length and number of stems. All rooted cuttings should be pinched, leaving behind three to four pairs of leaves.

Hypericums do best in well-drained, balanced soil with average fertility and a pH of 5.0 to 6.3.

Plant out established rooted cuttings at a finished spacing of about 15 to 18 in. from one plant to the next, going down a row. Final spacing may vary, depending on whether the crop is grown in the field or greenhouse and how mature the crop is.

Growing Environment

Both Excellent Flair and Autumn Blaze thrive under a moderate growing environment, with night temperatures at 50° F (10° C) and days 75° to 85° F (24° to 29° C). By growing under these conditions and using mum lighting to extend daylength, it may be possible to achieve two flower flushes per year. Dutch research shows that stem quality, length, and production are improved when plants are given a total of fourteen hours of light (normal and artificial combined) per day until average stem length is about 12 to 15 in. long. At this time, increase total daylength to twenty total hours and maintain it until flowering is completed. Then, reduce daylength to fourteen hours until berry maturity or harvest.

Normally, though, when only natural light levels are available, the crop produces one quality flush and, consequently, only one quality berry setting per year. One-year-old plants from rooted cuttings yield two to three stems the first year, six to eight the second year, and twelve to sixteen thereafter.

Nutrition

Good culture practices also include managing crop fertility, especially nitrogen. Before flowering occurs, a minimal amount of nitrogen encourages flowering. After the berries have fully developed, back off on excessive watering so as not to burst them.

Pests

H. androsaemum and *H. inodorum* are both susceptible to mildew and rust fungi if conditions are mild and wet or humid. Follow a proper fungicide control program. Drip irrigation versus overhead watering is also beneficial in controlling disease.

Postharvest

After the berries have colored, it's time to harvest. Leave about a half inch of stem behind after harvest. Postharvest treatment, using a bactericide designed to help products with semi-woody stems take up water, will greatly increase the vase life.

Jeff McGrew, Horticulture Products and Services, Mount Vernon, Washington. April 1998.

Impatiens

❀

Grower's Guide to New Guinea Impatiens

Jack Williams

As consumer demand for New Guinea impatiens grows, so does your need for good cultural tips for a successful crop. Here's your guide to starting and establishing a high-quality crop.

Propagation

New Guinea impatiens are relatively easy plants to propagate. The key to successful propagation is to maintain appropriate cultural conditions. Using rooting hormones is optional because most New Guinea impatiens cultivars root without them. Most New Guinea impatiens are propagated during late fall and early winter when light intensities are low enough that additional shade isn't required. Some tips for success:

- Start with good-quality cuttings from a reliable supplier.
- Provide media temperatures of 70° to 75° F (21° to 24° C).
- Provide air temperatures of 68° to 72° F (20° to 22° C) night and 70° to 80° F (21° to 27° C) day.
- Mist application for New Guinea impatiens should provide enough moisture to prevent heavy wilting of cuttings, yet allow cuttings to stay dry enough to avoid rot. Treat these plants like geranium cuttings when setting mist clocks. The most common cause of rot in young New Guinea impatiens cuttings is overmisting, which will result in botrytis and other fungal diseases.

Troubleshooting during Propagation

Problem: Cuttings rot in propagation.

Potential cause: Disease: Botrytis can result from damaged cuttings or cuttings with open flowers. Rhizoctonia may be the result of contaminated rooting media. Cultural: Excess mist application.

Problem: Cuttings fail to develop roots.

Potential cause: Cultural: Rooting media temperature is too cold. This may be aggravated by excess mist application.

Establishing Rooted Cuttings

You can grow New Guinea impatiens in a variety of forms, containers, and growing media. Always consider the importance of water-holding capacity when selecting a soil mix, as these plants require more water as they grow into their finished size and are sold for use by the general public. During the early establishing stage, avoid excess moisture, as this will cool media and slow root development. If there is a "trick" to getting New Guinea impatiens off to a good start, it's to keep greenhouses warm and monitor soluble salts carefully.

Temperatures for establishing this crop are similar to those used for propagation. Maintain night temperatures of 68° to 72° F (20° to 22° C) and day temperatures of 70° to 80° F (21° to 27° C) until active growth is evident. These temperatures encourage development of leaves and vegetative growth.

Also, avoid fertilizing young plants until two to three weeks after transplanting. After cuttings develop good roots and vegetative growth is evident on the upper plant, begin fertilization with complete and balanced fertilizers. Total applied nitrogen should average 100 to 150 ppm constant liquid feed at this time. Avoid fertilizer mixtures with elevated minor elements.

Troubleshooting during Establishment

Problem: Rooted cuttings fail to develop new roots and appear stunted after several weeks.

Potential cause: Disease: Look for signs of damage on root systems resulting from pythium or other fungal pathogens. Insects: Look for fungus gnat larvae that attack young root systems and can vector diseases. Cultural: Problem can be caused by excess moisture in soil, by high soluble salts if fertilizer has been applied either as constant liquid feed or as constant release, or by low temperatures.

Problem: Rooted cuttings develop symptoms of black streaks in the main stem; leaves may show concentric ring spots. Leaf tips curl, and marginal burn is evident.

Potential cause: Disease: Can be caused by virus infection by either TSWV or INSV. Isolate plants and have samples tested by a qualified crop diagnostic laboratory. Botrytis can also cause some symptoms that mimic one of the virus diseases. Look for patterns of moisture on leaves that may be responsible. Insects:

Thrips are the only known vectors of virus transmission. Check New Guinea impatiens and adjacent crops for the pest. Cultural: Phytotoxicity may occur from excessive fertilization or pesticide applications.

Jack Williams, Paul Ecke Ranch, Encinitas, California. February 1997.

Program Helps Ensure Disease-free Vegetative Impatiens

Gary W. Moorman

Until recently, impatiens propagators have had no procedures to guarantee healthy stock plants for vegetative production. That's all changed with a new grower certification program from Pennsylvania State University and the Pennsylvania Department of Agriculture. The program sets parameters for growers to meet so that buyers know plant material is free of specific viruses when purchased.

Because of impatiens' importance in the greenhouse industry and the damage to the crop in recent years by western flower thrips and impatiens necrotic spot virus, Pennsylvania State University, in cooperation with the Pennsylvania Department of Agriculture, has developed protocol to produce impatiens "apparently" free of disease. Pennsylvania growers can participate in the certification process if they can meet the criteria.

Pennsylvania has the authority to establish legal rules pertaining to the production and certification of plants. A similar program was developed many years ago for geranium production. Certification occurs only if the pathogens of concern aren't detected and none of the plants exhibit symptoms. Note: Certified plants are designated "apparently" free of those specific pathogens but may not be free of all pathogens.

Setting Parameters

The first criterion participants have to meet is having a source of impatiens that has been indexed or tested and found apparently free of certain viruses. When growers have those plants, they grow them on in a greenhouse isolated from all other plants. The new plants are inspected on a regular basis for disease symptoms and the presence of mites and insects.

Initially, 2% of new plants that will become the grower's stock are randomly tested for impatiens necrotic spot virus, tomato spotted wilt virus, and tobacco mosaic virus. Indicator plants are also inoculated with sap from these plants as a test for the presence of other viruses.

Following the initial sampling, any plants with suspicious symptoms are tested. This protocol must be followed continuously from the time new stock plants enter greenhouses until all cuttings and stock plants are sold or discarded. Growers must obtain new stock each growing season—they can't carry over any impatiens.

The First Step

One operation, Vic's Plant Place Inc., Lancaster, Pennsylvania, owned by J. Victor Vanik, has been participating in the Pennsylvania program with rosebud impatiens.

A pilot project ran for two years before the certification process for the 1995–96 growing season officially began. During the pilot project, researchers studied impatiens growing procedures, watched how plants were moved from the stock house to the rooting area, and documented steps along the way to determine what needed to be changed or what could be kept the same to ensure that the impatiens remained healthy. They also identified scouting and testing procedures to verify plants' health.

In addition, employees at Vic's Plant Place were trained to understand the importance of their work for the overall objective of producing healthy plants. This also encouraged employees to assist in scouting for plants with suspicious symptoms.

Making It Work

The initial plant material used as stock is received from D.S. Cole Growers, Loudon, New Hampshire, in October. That material originates from cuttings from Ball FloraPlant, West Chicago, Illinois, where it's tested for particular pathogens.

Rosebud impatiens production at Vic's Plant Place is complete by the middle of March, and stock plants are then sold or discarded. New stock plants are established each October. By starting with healthy plants and verifying their continued health with Pennsylvania Department of Agriculture testing and scouting and Pennsylvania State University scouting, the rooted cuttings produced are then officially certified by the Pennsylvania Department of Agriculture to be "apparently" free of pathogens for which they were tested.

Gary W. Moorman, professor of plant pathology, Pennsylvania State University, University Park, Pennsylvania. February 1997.

Ivy

And You Thought It Was Just an Ivy

Pat Hammer

Many vining plants out there are parading around as ivy, such as Boston ivy, Swedish ivy, German ivy, grape ivy, Kenilworth ivy, and poison ivy. Actually, they aren't true ivies at all. English ivy belongs to the genus *Hedera*, which includes several species. The ivy most often grown is *Hedera helix*, or English ivy. The other more common species are: *H. canariensis*, Canary ivy, or Algerian ivy (the freeway ivy in California); *H. colchica*, Persian ivy; *H. nepalensis*, Nepal ivy; *H. rhombea*; and *H. pastuchovii*,

Russian ivy. *Hedera* does include other species, but they're usually only found in the hands of ivy collectors.

English ivy can be found in every florist shop, garden center, and retail greenhouse or nursery. In the Florida foliage trade, it's grown more than any other foliage plant. More than four hundred cultivars of English ivy are grown in the U.S., but only about thirty different cultivars are produced commercially.

Like so many other plants with a large number of cultivars, there's considerable confusion with names. Sometimes the mix-up with names comes from handwritten labels that are hard to read, or an alert grower finds an unnamed ivy in the supermarket and gives it a name. From time to time, ivies sport and send out significantly different shoots. If greenhouse workers propagate these shoots by mistake, you'll have the wrong name on a brand-new ivy, or the workers will have put a new name on a sport that already has a name.

Ivy is easy to propagate and easy to grow. Its requirements are minimal, and it will try to adjust to most growing conditions as long as they're fairly consistent. Although English ivy cultivars can vary slightly in their cultural requirements, they do better if they're grown in bright shade with consistent soil moisture and 55° to 60° F (13° to 16° C) nights. They can be watered frequently if grown in a lightweight, well-drained soil mixture. Almost every commercial grower has his or her own mix, but generally they're on the acidic side. One example is a mix California growers use. It consists of ⅓ peat, ⅓ perlite, and ⅓ redwood sawdust. With this mixture, lime is added to raise the pH to 6 to 6.5.

Consumers have become very sophisticated about ivy. They've learned to use it as a ground cover that replaces high-maintenance and time-consuming lawns. Homeowners are looking for the new, exciting variegated and gold ivies that add interest.

Ivy topiary has created a demand for smaller cultivars that accent artistic shapes. Gardeners have learned to use ivies as companion plants with spring, summer, and fall bulbs. They make a nice foliage cover when bulbs aren't in season. And ivy has always worked as a companion for window boxes, planters, and mixed hanging baskets.

English ivy remains the most elegant cascading element used in plant combinations. It softens container edges and improves their proportions.

With the increased demand within the foliage industry for new and exciting plants, it's important for ivy growers to keep up on new cultivars being introduced every year. The American Ivy Society has a testing program designed to study new ivies and get them into the hands of growers who want to stay one step ahead of the market. For more information about ivies, contact the American Ivy Society, P.O. Box 2123, Naples, Florida, 34106-2123, or visit them at: www.ivy.org.

Pat Hammer, president, American Ivy Society, and owner, Samia Rose Topiary, Encinitas, California. September 1997.

Kalanchoe

Kalanchoe

Jeffrey R. Watt

The kalanchoe is a flowering succulent that provides consumers with a colorful, long-lasting plant requiring very little maintenance. For growers, kalanchoes are a good crop to produce because of their capacity to be marketed for year-round sales. Although they are the leading potted crop sold in Europe, kalanchoes are not as popular in the United States yet, but they are increasing in popularity. They can be marketed for several uses—for example:

Centerpiece

Pot sizes ranging from 3-in. to 6-in. are excellent for centerpieces with the uniform height and great blooming power of newer varieties.

Dish gardens

Kalanchoes are great for dish gardens because their foliage and flowers show well with other plants used in arrangements.

In the garden

Kalanchoes perform well in the garden and are a great landscaping choice. Most varieties do well as an annual, and the foliage is attractive, even after the plant has finished flowering.

Large pots

Grown or transplanted in large containers, kalanchoes can be enjoyed as an exceptional patio plant.

Novelty/gifts

Smaller pot sizes, especially 3-in. pots, are excellent for small gifts and favors. Recent hybridizing of kalanchoe varieties has produced plants with several growth habits and a wide color range. Many newer varieties are characterized by strong branching, compact growth, and an abundance of large flowers, lasting five to seven weeks in the home if purchased at the correct stage. For consumers, kalachoes will do well in full light or in shade, and there is no need to fertilize, just occasional watering when the medium has dried out. For growers, they present relatively few problems with pests or diseases and have proven durable in shipping.

Propagation

Kalanchoes are propagated by vegetative cuttings. They should be rooted in a larger cell tray, 72-cell-size or larger, and should be misted or under "tents." The cuttings will root in ten to fourteen days, which is when the mist/tents should be taken off. After initial rooting occurs, the plants can be fed at a rate of 200-70-200 ppm until transplanting. Cuttings can also be directly stuck into 4-in. and 6-in. pots using the same misting procedures as in propagating flats. A license is required for propagating patented varieties.

Cultural Recommendations for Kalanchoes

Growing medium

We recommend 60% peat moss and 40% perlite or Styrofoam (or any well-draining commercial mix.)

Temperature

A temperature of 68° F (20° C) is optimal for initial rooting; 65° to 68° F (18° to 20° C) during flower initiation and growing.

Feed

We recommend constant feed of 200-100-200 during vegetative growth, clear water every other time during short-day treatment, then constant feed until two weeks before sales. Use clear water to finish.

Short-day treatment

Kalanchoes require six weeks of short days (at least fourteen hours of darkness) to initiate flowers. Short-day treatment is not needed after this initial six weeks. Use blackcloth from February 10 through October 20.

Lighting

Twenty foot-candles of light are recommended for one to two hours in the middle of the night from September 15 through March 21 to keep plants vegetative during the initial growth.

General Growing

No pinch is necessary for plants started in March through October. If HID lights are not available, from November through February make a soft pinch removing just the tip at the time of the start of short-day treatment. Generally, one week of long days in the summer and three weeks of long days in the winter are sufficient after transplanting a liner to get good vegetative growth. For six-in. pots, using three liners per pot, use two to four weeks of long days. Add three to four weeks for a direct stick program. As a guide for estimating final size of the plant, three more vegetative nodes will develop after start of short-day treatment before the flower stem develops.

Growth Regulators

For 4-in. production in winter, use B-9 at 2,500 ppm for four or five weeks after the start of short-day treatment. In the summer, apply three weeks, five weeks, and possibly seven weeks after the start of short-day treatment. For 6-in. production in winter, apply five weeks after the start of short-day treatment, and in summer, three weeks, five weeks, and possibly seven weeks after the start of short-day treatment. These are general guidelines that may differ by region and grower preference on the size of the finished pot. Kalanchoes do respond to negative "DIF" (or "inverted night temperatures").

Pests and Diseases

Maintain good ventilation practices to help control botrytis and powdery mildew. Many fungicides can be used, but do not use Karathane or Rubigan. Insects can be controlled on a see-and-treat basis.

Blackcloth Suggestion

The blackcloth system does not need to be anything elaborate. Simply pull 4-mil black poly over the plants. Make sure the plastic is wide enough to drop just beyond the sides of the bed. When the days are hot and bright pull the plastic at 7 P.M. and remove at 9 A.M. Since there is no need to blackcloth after the end of six weeks of short-day treatment, there will be no physical damage to the flower buds. Also, the plats can remain pot tight in trays during this treatment so they don't fall over while pulling the plastic on and off.

Jeffrey Watt is the sales and marketing manager for Mikkelsens, Inc., Ashtabula, Ohio. October 1994.

Kale, Ornamental

Flowering Kale: Fall Color for Late Sales

Bonnie Marquardt and Ron Schlemmer

Producing color long after most annuals, flowering kale is perfect for extending fall sales. Especially ideal for the fall landscape, flowering kale and cabbage both give outstanding color and long-lasting impact in the garden.

Flowering kale and cabbage differ in their basic leaf shapes. Cabbage has rounded leaves; kale has fringed leaves. Additionally, feather-leafed kales are differentiated by finely serrated or deeply notched leaves. They're taller than fringed kale or rounded cabbage, making them suitable for the center or as backdrops. All varieties are edible but slightly more bitter than the green vegetable kales. They're often used as salad garnishes.

Culture Tips

Growing flowering kale means treating it like late cabbage. Sow seed in June or July, and set plants out in July or August. Plants will take eight to eleven weeks, depending on temperatures, to produce a coloring plant. Growing times are directly related to cultivar, pot size, starter plant, and cultural environment. It's an excellent product for 6-in. or 1-gal. containers and can also be grown in 4-in. pots during the cool season.

Temperature

Temperature is the most important factor for coloring of kale. Plants begin to color on the central leaves when the average temperature is 60° F (16° C). The optimum temperature range for flowering kale is 50° to 70° F (10° to 21° C). Established plants can take temperatures as low as 20° to 25° F (-7° to -4° C) without damage. The cooler plants are grown, the more compact they will be. If seed is sown too early, when temperatures aren't cool enough to induce coloring, plants will continue to produce layers of green leaves until temperatures are cool enough for color development. This will produce a tall, columnar plant. After plants have colored, they're vernalized and will begin to bolt if temperatures warm up. Just a few days of unusually warm temperatures can make kale bolt.

Moisture
Grow on the dry side to prevent tall succulent plants.

Light
High light (full sun to part shade).

Fertilizing
Flowering kale and cabbage like a low, constant feed of a complete fertilizer. If you prefer a slow release fertilizer, you can also use it at a low rate.

Lack of nitrogen is a common problem in growing kale. It will cause lower leaves to turn yellow and drop. Maintain fertilization until temperatures drop, when you should see coloration begin. Grow plants with a constant feed program of about 150 ppm nitrogen from start to finish. Once coloring begins, too much nitrogen will produce green blotching on colored leaves.

Height control
Excessive stem elongation due to high temperatures is a major problem with growing flowering cabbage and kale during the summer. B-Nine and Bonzi have proven effective for reducing stem elongation.

Insect/disease problems
Flowering cabbage and kale can be attacked by insects including aphids, cabbage looper, and leaf roller. When plants are young they can be susceptible to downy mildew if the soil stays too moist.

Marketing Ideas
Recently the kale market has roller-coastered, with some high peaks and low points. Remember that education is the key—growing it properly, then telling consumers how to use it. Work with landscapers on big plantings in busy traffic areas (parks, city hall, the library, the mall, etc.). Show kale planted side by side in masses or in mixed plantings. Mix color combinations using plants with red, pink, and white flowers (purple also blends nicely). Color bowls and additional containers also give a great show.

Flowering cabbage and kale are attractive complements to garden mum sales, offering a way to extend the fall season. While most flowering plants lose their flowers or color after several frosts, flowering kale and cabbage intensify in color and can last until past the first snowfall. Plant flowering kale and cabbage with other fall

annuals. Remembering texture, an assortment of different plant species gives defini-
tion to the garden. Plant them with species such as dianthus, stock, snapdragon,
primrose, pansy, and viola.

*Bonnie Marquardt, flower sales manager, and Ron Schlemmer, flower trial manager/breeder, American
Takii Inc., Salinas, California. July 1996.*

Kale Responds to Bonzi

In research at Uniroyal Chemical Company, flowering kale varieties treated with
Bonzi showed delayed bolting, a typical problem in warmer temperatures, according
to *Ornamentals Report*. Researchers say this may be due to less stress on plants. Some
varieties also displayed darker, more intense colors after treatment.

Plants respond best to summer-applied 10 ppm Bonzi drenches at labeled rates.
Researchers experimented with 4- and 6-in. pots.

Bonzi treatments on 6-in. flowering kale pots*

Rate	W. Pigeon	R. Pigeon	W. Kamome	R. Chidori	Average reduction
10 ppm	4.5 in.	3.75 in.	2.25 in.	4 in.	42.7%
20 ppm	3.75	3.5	2.75	3.75	46.5%
30 ppm	3.25	3.5	2	3.5	52.3%
40 ppm	2.25	3.5	2.5	3	56.2%
check	6.5	7.25	7	5	0%

*Measurements from pot rim to top leaf. Data collected 25 days after treatment.

Bonzi treatments on 4-in. flowering kale pots*

Rate	W. Kamome	R. Chidori	P. Beauty	W. Peacock	Average reduction
10 ppm	3.5 in.	3.5 in.	4 in.	3.5 in.	49.7%
20 ppm	3	3	3.5	3.75	53.8%
30 ppm	3	2.75	3	2.5	58.4%
40 ppm	3	2.75	3.5	2.5	58.8%
check	5	4.5	6.5	9.5	0%

*Measurements from pot rim to top leaf. Data collected 25 days after treatment.

Culture Notes, August 1997.

Lilies

❀

Height Control Possibilities for L.A. Hybrid Lilies

William B. Miller, Jason P. Ball, Garry Legnani, and Beth Hardin

L.A. hybrid lilies, a new group of lilies from interspecific crosses of *Lilium longiflorum* and Asiatic hybrid lilies, have great potential as pot plants with the development of good height control techniques.

This group of lilies has only been available for the last four or five years, and they show many of the best qualities of Easter lilies (good foliage, vigor) combined with the color possibilities of Asiatic hybrids. Other outstanding features are large, substantial buds, larger flowers than typically seen in Asiatics, and a more open flower than Easter lilies.

In the Netherlands, most L.A. hybrids are used for cut flower production. In the U.S., their increased use as a pot crop is dependent on effective height control. At Clemson University, we evaluated the effectiveness of growth regulator sprays, drenches and pre-plant bulb soaks to control L.A. hybrid lily height.

On February 15, we treated L.A. hybrid lily bulbs (5 to 5½ in. circumference, cultivar Royal Fantasy) with three growth regulators: A-Rest (33 ppm), Sumagic (5, 10 and 20 ppm), and Bonzi (100 and 200 ppm). We soaked five bulbs in 600 ml of each solution for five, ten and sixty minutes. Control bulbs were in distilled water for the same three time periods. After soaking, bulbs sat overnight at room temperature, then the next day we planted them in 6-in. standard pots in a standard soilless greenhouse mix (Fafard 360). We next placed the bulbs in the greenhouse and grew them with standard cultural practices, with night temperatures of 63° F (17° C) and days approximately 12° F higher.

We planted other plants on February 16 and placed them in the greenhouse. On March 12, we sprayed plants with A-Rest (33 ppm), Sumagic (5, 10, 20 and 30 ppm

Table 1. Effects of A-Rest and Sumagic sprays and drenches on final height and timing of Royal Fantasy L.A. hybrid lilies

Treatment	Number of flowers	Plant height at flowering[1]	Days to flower
Controls	3.2	16.3	74.7
Drenches			
Sumagic 0.1 mg/pot	3.5	14.2	73.8
Sumagic 0.2 mg/pot	3.8	10.2	74.5
Sumagic 0.3 mg/pot	4.0	8.0	75.0
A-Rest 0.5 mg/pot	3.3	10.5	74.5
One spray			
A-Rest 33 ppm	3.7	16.9	75.0
Sumagic 5 ppm	3.3	14.4	74.0
Sumagic 10 ppm	3.3	14.4	74.5
Sumagic 20 ppm	3.5	14.4	74.8
Sumagic 30 ppm	3.7	12.8	76.0
Two sprays			
A-Rest 33 ppm	3.8	14.2	74.2
Sumagic 5 ppm	2.8	12.9	75.5
Sumagic 10 ppm	3.2	12.3	75.2
Sumagic 20 ppm	3.5	9.6	76.0
Sumagic 30 ppm	3.5	10.1	78.3

[1] Plant height only. Add pot height for total product height.

once or twice) or drenched them with A-Rest (0.5 mg) or Sumagic (0.1, 0.2 or 0.3 mg per pot). Sprays were the standard rate of 2 qt. of growth regulator per 100 sq. ft. Plants averaged about 2 in. on the first spray date (March 12); the second spray date was fourteen days later. We drenched only once, on March 12.

Soaks

The bulb soaks were highly effective for controlling plant height (see graph). Pre-plant soaks in 33 ppm A-Rest reduced height by more than 50% relative to the controls, and increasing soak time didn't change the final height obtained. All Sumagic concentrations and soak lengths produced plants that were too short for 6-in. pots. Although we found no difference between five- and ten-minute soaks, plants soaked for sixty minutes were shorter for all Sumagic concentrations. The 100 ppm Bonzi soaks gave very nicely proportioned plants when soaked for five minutes.

Control plants flowered sixty-five days after planting. Relative to the controls, all growth regulator bulb dips caused an average flowering delay of five days, and except for the severe sixty-minute soak in 20 ppm Sumagic, when plants didn't flower, flowering delay didn't seem to be closely related to growth regulator dose. Growth regulator bulb soaks didn't appear to affect the number of flower buds or aborted buds, even on plants that were severely stunted in height. Overall, plants averaged 3.6 flowers per plant.

Drenches

Drenches were very effective for height control of Royal Fantasy (see Table 1). The standard treatment of 0.5 mg A-Rest reduced height by about 35% with no flower delay. Sumagic drenches from 0.1 to 0.3 mg per pot gave increasingly effective height control, again with no flower delay, with 0.2 mg being optimal.

Sprays

Overall, sprays were substantially less effective for height control than the bulb dips or drenches we used. Two 33 ppm A-Rest sprays caused only a 13% height reduction compared with the control. As expected, Sumagic sprays increased in effectiveness with increasing concentration, and two sprays were always much more effective than a single spray. For example, a single 10 ppm Sumagic spray reduced height by 12%, but two sprays reduced height by 20% compared with the controls with no delay in flowering.

Future Possibilities

With the reduction in height in this preliminary experiment, it appears that L.A. hybrid lilies can be tailored for production in 6-in. pots with no reduction in flower number or quality. The percentage of height reduction we saw with 33 ppm A-Rest soaks is somewhat greater than seen in Roy Larson's Easter lily bulb dipping work at North Carolina State University in the mid-1980s. It's safe to assume that the various L.A. hybrid lily cultivars will respond differently to bulb soaks, sprays or drenches. Although our results show very good effects, it's possible we worked with an especially sensitive cultivar for these experiments. Because we've looked at only one cultivar, we can't make absolute recommendations for height control of all L.A. hybrids.

However, a starting point for height control in this group of lilies might be two 10 ppm Sumagic sprays, a 0.1 to 0.2 mg Sumagic drench or a 0.25 to 0.5 mg A-Rest drench. Pre-plant bulb dips will likely need to be in the range of 10 to 20 ppm A-Rest, 1 to 3 ppm Sumagic and 100 to 200 ppm Bonzi for two to ten minutes.

L.A. hybrid lilies have other potential as well because the garden performance of these hybrids is excellent. Consider "green" sales with five to nine bulbs in large containers twelve inches or more diameter. Items such as these would need no growth regulators and would give consumers long-term value as buds develop and flower.

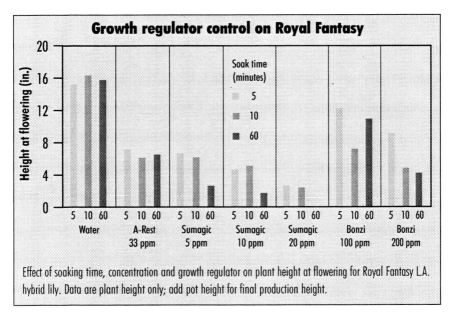

Growth regulator control on Royal Fantasy

Effect of soaking time, concentration and growth regulator on plant height at flowering for Royal Fantasy L.A. hybrid lily. Data are plant height only; add pot height for final production height.

Bill Miller is associate professor; Jason Ball is an undergraduate student; Garry Legnani is a graduate student; and Beth Hardin is research associate, Department of Horticulture, Clemson University, Clemson, South Carolina. The authors thank Mike Hollander, Sheppard West, Eugene, Oregon, for donating bulbs for this work. The bulb soak portions of this project were conducted by Jason Ball, whose presentation of the work won first place in the undergraduate competition of the 1996 Southern Nurserymen's Association Research Conference. March 1997.

Crop Culture: Lilies—Asiatic and Oriental

William B. Miller, Robert Miller, and Bob Miller

The major advantages of hybrid lilies are relative ease of production, high crop value per square foot, an ever-increasing variety selection, better height control possibilities through chemicals and genetics, excellent value for the consumer and, in the case of Asiatic hybrids, low greenhouse temperature requirement. Here are some tips for growing these popular spring crops.

Media

Use good, well-drained media with high air-filled porosity. Maintain pH at 6 to 6.5; lower pH can lead to fluoride leaf scorch. Avoid perlite and superphosphate in hybrid lily media mixes. Keep in mind that mixes containing pine bark tie up growth regulators, reducing their efficacy. Typically, three bulbs (12/14 size) are planted per 6-in. pot, five bulbs per 8-in. pot. Water in very well.

Fertilization

Fertility programs vary widely. Easter lilies finish taller when grown hungry, and hybrids probably do as well. The basic program at Clemson University is to use 200 ppm nitrogen at each irrigation with clear water applications on weekends. This approach has never resulted in salt problems, yet produces good foliage color. Because there is low alkalinity water in South Carolina, we give at least one application of calcium/potassium nitrate at 200 ppm nitrogen per week, and use 20-10-20 peat-lite for other irrigations. Specifics of

fertilizers haven't been intensively studied for hybrid lilies. We also like to let the crop run on the dry side for the first two to three weeks or until stems are one to three inches tall, as we think this improves stem root growth.

Temperatures and Timing

Asiatics need to be grown cooler than Orientals. Grow Asiatics at 55° to 60° F (13° to 16° C) nights, and 70° F (21° C) or less days, no more than 85° F (29° C) days. Orientals respond better to 62° to 65° F (17° to 18° C) nights and 75° F (24° C) days. As with Easter lilies, timing is mainly controlled by twenty-four-hour average temperature, and some level of height control can be achieved by using negative DIF (warmer nights than days).

Always keep in mind the formula for calculating average temperature, because the average temperature isn't simply the average of the day and night temperatures. Calculate average temperature as follows:

[(Day temp x hours of day temp) + (night temp x hours of night temp)] / 24

As a general rule, the number of days from planting to emergence is based on the length of cold storage or freezing in. Days to emergence varies from two to three weeks for early crops to three to four days for crops planted in May and later. Beyond that, timing is highly cultivar- and season-dependent. For example, timing to visible bud may vary from two to four weeks after emergence. The number of days from visible bud to flowering is also highly variety-specific and is a long period for most Orientals.

Lighting

Sans Souci, an older Oriental cultivar, can be forced three to four weeks faster through the use of long days (mum lighting). Light for four hours from 10 P.M. to

2 A.M. with at least 10 f.c. (60 watt bulbs, 4 ft. apart, 4 ft. above the benches). Stargazer responds in a similar manner, but extensive studies with other Oriental hybrid cultivars haven't been reported, although you should see some positive timing response.

Height Control

A-Rest

A-Rest drenches are the most common means of chemical height control. Drench rates are typically in the 0.25 to 0.5 mg/pot range. Early application as soon as roots are growing is a key to good chemical height control. When using multiple bulbs per pot, apply the drench when the first two shoots are ½ in. tall. Don't wait for all shoots to emerge. Split applications are usually better than a single application, as the first application (at ½ in.) can help reduce early stretch, and the second (seven to fourteen days later) is more readily absorbed due to the better mass of stem roots present. Again, pine bark mixes tie up growth regulators and reduce their effectiveness.

Sumagic

A Sumagic drench is also highly effective and has the advantage of substantial cost savings compared to A-Rest. Clemson University researchers trialed Sumagic at 0 to 0.3 mg. per pot on eleven hybrid cultivars. *These high rates are above those recommended on the label, and the plants were growing in pine bark media.* The treatments had good height control with no effect on days to visible bud or number of flowers and no flowering delay.

Cold Storage

As with Easter lilies, there is the temptation to cold-store budded plants before sale, but Oriental hybrid lilies store much more poorly than Easter lilies. Upon removal from the cooler, plants often drop a large portion of their leaves, and quality is drastically reduced.

Recent research at Clemson University has indicated that using lights in the cooler can help with this problem. Researchers held Stargazer plants in darkness or under two 4-ft. fluorescent lamps about 4 ft. above plants in a 40° F (4° C) cooler for seventeen days. They saw a very beneficial effect of lights. While this is a starting point, more work needs to be done on postharvest storage of hybrid lilies.

Troubles

Bud blasting

Caused by low-light forcing or ethylene in the greenhouse atmosphere, bud abscission can happen during postharvest transit and is exacerbated by ethylene and high temperatures.

Root rot

Rot is prevented through a regular fungicide program alternating materials effective against pythium, rhizoctonia and fusarium. Within one to three days of planting, drench the crop using this schedule:

Table 1. Fungicide drench program for root rot control in hybrid lilies

Potting: Terraclor (4 oz./100 gal.) and Subdue (½ oz./100 gal.) Monthly intervals: Use a mixture of one each from Group 1 *and* Group 2 below. During alternate months you can use Banrot 40 WP by itself at 6 to 12 oz./100 gal. Banrot is effective against both rhizoctonia and pythium.			
Group 1 (pythium control)		**Group 2 (rhizoctonia and fusarium control)**	
Banol 65 EC	20 oz./100 gal.	Cleary's 3336-F	1 1/2 pt./100 gal.
Truban 30 WP	3 to 10 oz./100 gal.		
Truban 25 EC	4 to 8 oz./100 gal.		
Terrazole 35 WP	3 to 10 oz./100 gal.		
Terrazole 25 EC	4 to 8 oz./100 gal.		

William B. Miller, Clemson University, Clemson, South Carolina; Robert Miller and Bob Miller, Dahlstrom and Watt Bulb Farms, Smith River, California. December 1996.

Handling Easter Lily Bulbs

William B. Miller

To ensure that you produce a successful crop on time for Easter, try these tips for early handling of your bulbs.

Easter lilies require six weeks of low temperatures to flower for Easter. This temperature treatment hastens emergence, time to visible bud and flower. It also improves uniformity of the entire crop and reduces final plant height. On the other hand, flower numbers are reduced with increasing weeks of cold. Bulbs can be cooled in cases by the grower or cooled in the pot with CTF (controlled temperature forcing) or natural cooling techniques after planting.

Important Facts to Remember:

Inspect bulbs upon arrival, noting damage to cases. Also, open boxes and check for evidence of freezing. While Easter lilies can withstand temperatures down to 28° F (-2° C), temperatures below this can be harmful. Report any problems to your supplier.

Never let lily bulbs dry out. Unlike many spring bulbs, lilies don't have a papery skin to prevent bulb water loss. If you're doing your own case cooling, your non-cooled bulbs should arrive in late October. Place in the cooler immediately. Be sure

to maintain case temperature at 35° to 40° F (2° to 4° C) for Ace and 40° to 45° F (4° to 7° C) for Nellie White. Base your cooler settings on packing material temperature, not just air temperature. To perceive the cold temperature, bulbs must be moist but also exposed to plenty of fresh air. Since packing material can dry out during shipping and cooling, consider adding one to two quarts of cold water per case to maintain moisture level during cooling.

If you bought commercially precooled bulbs, they need to be planted upon arrival and placed in the greenhouse. In contrast, pot-cooled bulbs (CTF) arrive non-cooled and must be planted upon arrival and placed in a warm (63° F/17° C) location for one to three weeks of rooting out. To determine how long you can maintain 63° F (17° C), count back sixteen weeks from Easter. This is the date the crop should come into the greenhouse. Count back another six weeks, and this is the date you should move pots to the cooler. The difference from that date and planting is the 63° F (17° C) rooting period.

For example, for Easter, April 7, 1996:

Date into the greenhouse:	December 17
Date into the cooler:	November 5
Rooting at 63° F (17° C):	From planting date to November 5

Be sure to give the full six weeks of cooling! Skimping on cooling will increase forcing difficulty and reduce crop uniformity. Some growers still use natural cooling. In this case, non-cooled bulbs are potted upon arrival and placed in sheds or cold and exposed to prevailing cool temperatures of October, November and December. It's important that bulb temperatures of 35° to 45° F (2° to 7° C) be maintained as best as possible. Records should be kept on minimum and maximum temperatures and duration to provide a guide. The main advantage to natural cooling is its low cost; however, the lack of consistent control over temperatures can lead to problems.

Lily soil mixes need three important characteristics: excellent drainage, excellent water-holding capacity and excellent fertilizer-holding capacity (cation exchange capacity). Mixes with these characteristics help eliminate problems such as root rot, elevated soluble salts and nutrient deficiencies. Avoid superphosphate because it supplies fluoride that can cause leaf scorch, especially in Ace and when pH is lower than optimal. It's important to begin the crop with a growing medium that has the proper nutritional properties. Low nutrition early in the crop will reduce final plant quality, especially in terms of lower leaf senescence. Calcium is especially important for healthy root growth and to prevent leaf scorch.

After putting plants in the greenhouse, maintain the crop at 60° to 62° F (16° to 17° C) soil temperatures until emergence. For slow emerging crops, first try 2° to 3° F higher. Even soil temperatures as high as 70° F (21° C) should be okay as long as bulbs have been fully vernalized. Use insurance lighting *only* if bulbs are suspected of having less than six weeks of vernalization. It probably won't be needed this year due to the medium Easter date. Lighting is most effective within the two-week period at and just after emergence. Best results are obtained by using night break lighting from 10 P.M. to 2 A.M. You don't need to sort emergence groups for insurance light application.

William B. Miller, Clemson University, Clemson, South Carolina. October 1995.

Limonium

Limonium

Daijiro Harada

Limonium's unique branches are essential as filler for bouquets, corsages, baskets, and other flower arrangements. Plants for commercial production originate from tissue culture produced in Japan and Taiwan. The advantages of tissue-cultured plants are uniform growth, longer stem length, better flower color, and longer vase life. Other advantages include disease-free plant material and much higher crop yield.

While different limoniums look very similar, growers should recognize that there are differences in plants bred from different genera or parents. Growing different limonium varieties under the same cultural conditions and crop management may cause problems.

Basically, limonium can be divided into two types, each requiring different handling. Type I consists of seasonal flowering limoniums. This group flowers mainly in early to late summer, depending on the variety. Plants must be well established and exposed to cool temperatures to initiate flowers. If the plants aren't established, they won't produce flowers, even under low temperatures. Also, even if plants are mature, without a cold period they continue growing vegetatively. Plants planted in spring to summer develop sufficiently to withstand the cool temperature treatment for flowering the following summer.

Type II limoniums are free-flowering. This group has growth that is temperature-dependent. Varieties in this group flower throughout the year if temperature and light are appropriate.

Temperature and Light

Although growing temperature ranges differ among varieties, limonium is quite comfortable with 68° to 77° F (20° to 25° C). Type II, free-flowering varieties, flower throughout the year depending on temperature and light.

High light intensity may be substituted for some varieties, such as Saint Pierre or Beltlaard. These varieties will flower in winter with high light intensity even if night temperatures do not reach the required minimum temperature. These varieties usually need a minimum temperature of 60° to 65° F (16° to 18° C) for flowering in winter. Type I, seasonal flowering limoniums, require a cold period, or natural winter, to induce flowering.

Soil

Sandy or sandy clay soil is preferred, but any well-drained soil is suitable. Good drainage is essential for limonium cultivation. Soil pH of 6.5 and an EC of 0.5 are ideal.

Fertilization

Limonium varieties don't require much nutrition. Moreover, excess fertilization is not only costly but may result in unnecessarily tall crop height, weak stems, and flower abortion. Following is a guide for basic fertilizing for Type I, seasonal-flowering limonium: 34.5 oz. each of N, P, and K per sq. yd. Mixed fertilizer or slow-release fertilizer: 553 to 737 lbs. of organic manure per 100 sq. yd., depending on soil conditions. For Type II, free-flowering hybrids, such as Saint Pierre and Beltlaard, use N:P:K at a ratio of 2:1.6:3 as basic fertilizer. Try compound fertilizer 14-12-9 at 14.75 lbs. per 100 sq. yd. Use potassium sulfate, 3.7 lbs. per 100 sq. yd.

Watering

Generally, a good water supply is important to establish plants during vegetative growth. Once plants begin producing flower stems, sharply reduce water. Mature limoniums require very little water, especially free-flowering hybrid limoniums that will grow tall. Excess watering elongates stems and makes them very weak. The exception to this is the Emille family and Charm Blue, which take much more water. Freely watering these varieties helps the flower stems grow taller.

Harvest

Greenhouse cultivation is recommended for the highest quality stems and continuous production of free-flowering types. Yet all limoniums can be grown outdoors, except for a few tender perennials that you must keep over the minimum temperature required. Covering is essential to protect flowers from rain damage at harvest. Most varieties are ready to harvest when 70 to 80% of the flowers on the stem are open.

Daijiro Harada, Miyoshi & Co. Ltd., Yamanashi, Japan, in the 16th Ball RedBook. *April 1998.*

Lisianthus

Cut Flower Lisianthus

Takaai Miura

Lisianthus *Eustoma grandiflorum* is a native of North America, known to grow wild in the great plains of Texas and parts of Colorado, and commonly called prairie gentian. Original colors are mostly blue, pink, and ivory white.

Very little breeding work has been done until recent years for improving lisianthus for cut flower culture purposes. It was not until the introduction of F_1 hybrids that they became highly popular as a cut flower item throughout the world. Subsequently, there are two main growing seasons for cut flower production: winter to spring sowing for summer to fall cutting, and summer to fall sowing for midwinter to early spring cut flower harvesting.

In addition to the introduction of F_1 hybrids, mechanical sowing machines and plug trays, as well as pelleted seed, greatly accelerated and encouraged cut flower production. Lisianthus seed is very small, about 625,000 seeds per ounce.

Within a relatively short period of time, several new hybrids, including late, medium, and early flowering groups as well as spray types and double flowering varieties in several new colors including picotees, were introduced to the trade. It is these new colors and types that have made lisianthus a favorite for retailers and consumers.

Growing Medium

Well-sterilized growing medium with adequate moisture-holding capacity with good aeration is recommended for maximum results. Jiffy Mixed is used by many professional growers with good success as it is sterile and neutral. Never grow lisianthus in the same soil successively. For high quality cut flower production, soil with abundant organic matter is recommended. Mix about two tons of organic matter per 10,000 sq. ft., adding 25 to 35 lbs. of NPK per 10,000 sq. ft. A pH of 6.5, E.C. 0.5 to 1.0 is most preferred.

Germination Temperature

The optimum germination temperature ranges from 65° to 70° F (18° to 21° C). After complete emergence grow at a minimum temperature of 60° to 65° F (16° to 18° C) and a maximum temperature of 75° to 78° F (24° to 26° C). Do not allow temperatures to exceed 85° F (29° C) for any length of time as this will result in temporary dormancy, forming rosettes instead of normal flower buds.

Culture

Fifteen to twenty days after sowing, when seedlings develop four nodes (eight true leaves), care should be taken not to allow seedlings to dry out. After this time, the plants can be transplanted to a permanent location, spacing seedlings 5 by 7½ in. apart or 6 by 6 in. apart. Keep seedlings saturated until they are permanently rooted. Once seedlings develop roots, reduce water and allow the root system to develop adequately. Until flower buds appear on plants provide sufficient moisture and fertilizer. After flower buds appear, gradually reduce water to allow full development of plants. Also, extra care should be taken during spraying operations so as not to injure young flower buds.

Light

Provide ample light. Never cover seed, and provide as much light as possible during emergence and during the entire growing cycle.

Pest and Disease Control

Botrytis and fusarium are the most common diseases. For high-quality cut flowers, spray regularly against thrips and leaf mites.

Sakata's Heidi and Flamenco series can be cut with four to five flowers per stem and Double Flowering Echo series with at least three flowers per stem.

Takaai Miura is flower sales manager at Sakata Seed America, Morgan Hill, California. June 1995.

Osteospermum

Quick Cultural Tips for Osteospermum

Anne Whealy

Osteospermum's (African daisy) versatility and weather tolerance make it an ideal choice for hanging baskets and containers. Unlike many varieties of osteospermum on the market today, Stardust *Osteospermum compositae*, an evergreen perennial, is a true garden variety. Introduced by Farplants Ltd., Littlehampton, United Kingdom,

its average height over five years is 2 to 2½ ft., with an average spread of 2 to 2½ ft. It flowers from April to October, with occasional flowers in mild winters until Christmas.

Flower and Foliage

Purple-pink, iridescent ray florets display a bluish purple undersurface overlaid with bronze. Stardust presents a central disk of tiny, light brown florets with blue (occasionally purple) tips opening bright yellow. Flowers open in daytime and close at dusk. Cut flowers remain open under domestic lighting. Flowering period is from April to October. Foliage is mid-green and lance-shaped with a slight sheen, lighter undersides and a stiff, up-pointed form. Leaves are aromatic. Light green overlaid with bronze, Stardust's stem is short, stiff and ascending and almost totally clothed in leaves. Flower stems are ribbed, solid and upright.

Growing Conditions

The original plant has been growing in Yorkshire since 1985 and has successfully survived severe conditions with temperatures below 50° F (10° C) without damage.

Plants in trials have shown little reaction to frost. Wind protection in severe conditions would be beneficial. It prefers full sun, but light shade is suitable. Stardust will grow anywhere but prefers south, west, and east, in that order.

Any reasonably drained soil is suitable. Stardust is very tolerant of dry soils, but waterlogged or poorly drained soils are unacceptable. Allow to dry thoroughly between waterings. Fertilize every other watering with a balanced soluble fertilizer at 250 to 350 ppm nitrogen. Slow-release fertilizers also produce good results. It doesn't need growth regulators if proper cultural practices are followed. A hard pinch after roots are established in the final container produces a premium plant. Pruning is not necessary. Simply remove dead flowers to encourage further flowering.

Garden Use

In borders and beds, osteospermum is best as individual plants, or in threes at 18-in. (45 cm) spacing, or as full-season (April to October) bedding plants at 12-in. (30 cm) spacing. It is excellent for hanging baskets, window boxes or patio tubs, where its tolerance of dry conditions is an added bonus. Its long, sturdy stems and extensive flower life make it an excellent cut flower.

Propagation is prohibited. Sources are Skagit Gardens, Mount Vernon, Washington, and Ball Seed Co., West Chicago, Illinois.

Anne Whealy, executive director, Proprietary Rights International, Roanoke, Texas. January 1996.

Controlling Flowering in Osteospermum

Osteospermum is fast gaining popularity as a colorful pot crop to fill the sales gap between Easter and the start of the bedding plant season. Native to the Cape region of South Africa, plants have daisy-like flowers and dark green lobed leaves on plants that are naturally unpredictable and bloom in spurts. To find out the best way to control these wild tendencies and produce well-shaped plants full of flowers in 6-in. pots, Ayumi Suzuki and Jim Metzger, The Ohio State University, Columbus, tested osteospermum's response to photoperiod changes and cold treatments. Here are their findings, as reported in *Ohio Floriculture.*

Photoperiod

Researchers first attempted to manipulate flowering by altering the day length. Because osteospermum flowers in its native habitat in midsummer, researchers reasoned that summer's long days promote flowering. However, they found that even extremely long eighteen-hour photoperiods had little effect on flowering.

Vernalization

Because osteospermum is a perennial, another possibility for influencing flowering is cool temperatures. To test this, researchers transplanted rooted cuttings of Nairobi and Lady into 6-in. pots. They grew the plants on in a warm 75° F (24° C) day, 65°F (18° C) night greenhouse. After four weeks, researchers pinched plants, leaving seven to eight leaves, then moved them to a cool 45° to 55° F (7° to 13° C) greenhouse. Cultivars were returned to the warm greenhouse in groups after two, four, or six weeks. Some plants were left in the warm and cool greenhouses as checks. Throughout the experiment, researchers monitored timing and number of flowers produced in each treatment.

While vernalization had little effect on flowering, and sometimes even delayed flowering in both cultivars, the number of flowers produced increased substantially when plants were subjected to cool treatments for four weeks or longer.

Effect of vernalization on osteospermum flowering

Cultivar	Weeks of vernalization	Average number of days to open flowers	Average number of flowers per plant
Nairobi	0 (check)	121	1
Nairobi	2	95	7
Nairobi	4	98	41
Nairobi	6	106	54
Nairobi	Continuous	120	58
Lady	0 (check)	102	7
Lady	2	90	18
Lady	4	98	36
Lady	6	101	45
Lady	Continuous	114	45

Height Control

Through growth regulator recommendations for osteospermum have been published the researchers found that in addition to promoting flowering, vernalization treatments, particularly those six weeks or longer, are also effective for controlling height. Negative DIF after vernalization provides additional height control regulation if necessary.

Culture Notes, December 1997.

Pansy

126

It's Time to Think Pansies

Will Healy

Pansies are *the* undisputed annual for winter and spring flowering. In the South, pansies are planted in the fall, with flowering starting as soon as the temperatures reach the low 50s. In the North, pansies are planted during the fall or early spring, with flowering in late spring. Begin planning your pansy program in July.

Pansy production scheduling

Crop time		Night temperature (F)	Moisture	Fertilizer	Fertilizer rate (ppm) formulation	Frequency	Soil pH	Soil EC (mmhos)	Growth regulators
Sowing to transplant for a 512 plug									
Stage 1	3-7 days	65-75	Wet	0	No ammonia	None	5.0-5.5	<0.50	None
Stage 2	7 days	62-75	Moist	50-75	14-0-14	1 week	5.5-5.8	<0.50	B-Nine
Stage 3	2-3 weeks	60-75	Medium to dry	100-150	20-10-20	Alternate	5.5-5.8	<0.75	B-Nine or Arest
Stage 4	1 week	55-65	Medium	100-150	14-0-14	2 weeks	5.5-5.8	<0.75	Arest or Bonzi
Transplant to flower in a 4-inch pot or 367 tray									
Fall	6-8 weeks	65.75	Medium to dry	100-150 / 100-200	20-10-20 / 14-0-14	Once / Weekly	<6.0	<1.0	Bonzi or Sumagic
Winter	8-10 weeks	55-65	Medium to dry	100-150 / 100-150	20-10-20 / 14-0-14	Once / Weekly	<6.0	<1.0	B-Nine

*Alternate between 20-10-20 and 14-0-14, feeding with each once a week

Seed Form

There are several seed treatments available to improve the number of usable seedlings. Although raw or untreated seed produces acceptable results in the winter, most growers use primed or pregerminated seed for summer sowing. When raw seed is used in the summer, germination occurs in waves, which makes it difficult to manage seedling development within the tray. Pregerminated seed germinates uniformly, which results in uniform seedling development within the tray. Pregerminated seed will yield more usable seedlings per tray. Seeding accuracy quickly becomes the limiting factor in producing full trays when using pregerminated seed.

Plug Size

During the warm season of the year, growers use plug sizes ranging from 512 to 288 to improve germination and ensure plant survival after planting. At transplant time a rule of thumb is "as the temperature increases, the plug size should also increase to ensure survival."

Germination Conditions

Pansy seed requires cool, wet conditions for optimum germination. The quickest way to promote waves of germination is to germinate the seed at warm temperatures and run the flats dry. Pansies require constant temperatures of 65° to 70° F (18° to 21° C) from sowing until the radicle (root) emerges. At temperatures above 70° F (21° C), waves of germination occur, while above 75° F (24° C), germination is inhibited. Most growers use chambers to accurately control the germination temperature during Stage 1.

The moisture level around the seed is critical for success. To maintain the critical moisture level around the seed, thoroughly wet the flat before sowing. Use a fog system to maintain the moisture level in the flat. Cover the seed with coarse vermiculite to create a small "tent" around the seed. The large pieces of vermiculite maintain high humidity around the seed without smothering it. If you use fine vermiculite, the vermiculite plates smother the seed and prevent germination.

Fertilization Program

For top quality pansies, growers use a fertilization program that follows a couple of rules:

1. Avoid urea containing fertilizers.
2. Use fertilizers containing ammonia (20-10-20) to promote soft growth, leaf expansion, and stem stretch.
3. Use fertilizers containing just nitrate (14-0-14) to promote root growth, compact habit and well-toned plants.

Alternate between 20-10-20 and 14-0-14, depending on how the crop is developing. The initial feed for both plugs and finished crops is 14-0-14. Once roots develop, alternate with 20-10-20 to expand the leaves. As the leaves start to expand, switch back to 14-0-14 to prevent excessive stretch. During the warm periods of the year, pansies require periodic supplemental boron (Solubor at 0.25 oz./100 gal. or Borax 0.5 oz./100 gal.) to prevent boron deficiency.

Growth Regulators

A-Rest, B-Nine, Bonzi, or Sumagic keeps plugs and finished plants compact. Excessive ammonia and wet soils promote stem stretch, which requires additional growth regulator treatments. When using ammonia fertilizers, remember to apply the growth regulator before fertilizing for better height control. Since growth regulators require 2+ days to become effective, apply the growth regulator as the plant starts reaching the allowable size, not after it is too tall. When developing a growth regulator program, growers use B-Nine and A-Rest during the plug stage and Bonzi and Sumagic on the finished crop. Apply only as much growth regulator as is needed to control growth for two to three weeks.

Disease Control

Under certain environmental conditions, pansies experience a variety of diseases. Understand the environmental conditions that promote the disease, then manage the environment to reduce disease problems. When using chemicals to control disease outbreaks, follow the label carefully.

Thielaviopsis

Although black roots are a symptom, don't panic if your pansies have black roots. Pansy roots are transparent when they are wet and appear black because of the surrounding soil color. If roots still look black after you've rinsed them with water, then your plants likely have thielaviopsis. Reusing flats and placing plants on contaminated soil will spread the disease. Wet soil with a pH of 6.2 or higher promotes thielaviopsis. When growing outdoors, use iron sulfate drenches to keep the soil pH below 6.0.

Leaf spots

There are several leaf spots that produce gray, black, blue, or maroon lesions on pansy foliage. Plants that are poorly fertilized, grown under non-optimum conditions or high humidity will have more leaf spot diseases. Keeping the foliage dry going into the night is the primary control method to minimize spread of foliar diseases. Maintaining a regular fertilization program to encourage strong growth will reduce disease infections.

Yellow growing points

This is a problem that was observed in 1992, and we had several reports in the summer of 1994. Although the symptoms look like boron deficiency, the growing point is cream colored, while in boron deficiency the growing point is green. The problem appears when plants are placed in a high temperature environment.

Root rots

When grown outdoors, growers periodically experience pythium or rhizoctonia infections when excessive rains occur. The continuously saturated soil promotes the spread of the diseases.

Will Healy is manager of Technical Services, PanAmerican Seed, West Chicago, Illinois. July 1995.

Perennials

Forcing Perennials—Unlocking Their Mysteries

"Have you got slow-starting perennials in the spring?" asks Dr. Art Cameron, Michigan State University. "Then maybe they need long days, and they definitely don't get that in the early spring when days are short."

Understanding photoperiodic response, chilling response, juvenility requirements, controlling plant height, propagation, and storage techniques are all a part of the perennial plant research program spearheaded by Art at MSU, East Lansing.

March 21 is the spring equinox, when days equal nights. Growers producing LD-requiring plants naturally in the spring will find flowering is delayed until sometime in April. To meet LD requirements, Art recommends lighting from 10 P.M. until 2 A.M. at an intensity greater than 10 f.c. Incandescent lighting will work, he says, but it tends to cause more stretch than other light sources.

Some plants also require chilling (vernalization) to promote flowering. Art says cool temperatures can be provided in a cold greenhouse or cool store. Maintaining temperatures of 40° to 45° F (4° to 7° C) is ideal. Be careful if using a cool greenhouse because in poorly ventilated structures temperatures can rise high enough during sunny days (even in winter) to cause plants to devernalize.

In general a plant won't respond to cold treatment or to photoperiod until it has ended its juvenile period, Art says. Immature plants take longer to flower. However, you can't judge whether a perennial has ended its juvenile phase simply by its age. Counting leaf nodes is a better way to estimate plant age, he says. "Sometimes small plants are as old as larger ones."

The influence of long days (LD) and cold on flowering of herbaceous perennials

Flowering response		
No response to LD No response to cold	LD beneficial No response to cold	LD required No response to cold
Aquilegia x *hybrida*—cultivars that don't require cold (Songbird series) *Primula veris* Pacific Giants	*Leucanthemum* x *superbum* Snowlady (if no SD)	*Asclepias tuberosa* *Campanula carpatica* Blue Clips (if no SD) *Coreopsis verticillata* Moonbeam *Hibiscus* x *hybrida* Disco Belle Mixed (without SD)
No response to LD Cold beneficial	LD beneficial Cold beneficial	LD required Cold beneficial
Armeria x *hybrida* Dwarf Ornament Mix *Armeria latifolia* *Delphinium elatum* Blue Mirror *Dianthus deltoides* Zing Rose *Leucanthemum* x *superbum* Snowcap (a day neutral clone) *Scabiosa caucasica* Butterfly Blue *Veronica spicata* Blue	*Echinacea purpurea* Bravado *Lobelia* x *speciosa* Compliment Scarlet *Platycodon grandiflorus* Sentimental Blue	*Gypsophila paniculata* Double Snowflake *Oenothera missouriensis* *Rudbeckia ulgida* Goldsturm
No response to LD Cold required	LD beneficial Cold required	LD required Cold required
Aster alpinus Goliath *Aquilegia* x *hybrida*—hybrids that require cold *Heuchera sanguinea* Bressingham Hybrids *Iberis sempervirens* Snowflake *Lewisia cotyledon* *Linum perenne* Sapphire *Veronica longifolia* Sunny Border Blue	*Astilbe arendsii* *Coreopsis grandiflora* Sunray *Gaillardia grandiflora* Goblin *Lavandula angustifolia* Munstead Dwarf *Salvia superba* Blue Queen	*Achillea filipendulina* Cloth of Gold *Asclepias tuberosa* (after SD) *Chrysanthemum coccineum* James Kelway *Lavandula angustifolia* Hidcote Blue *Physostegia virginiana* Alba

Artificial light given at greater than 10 f.c. for four hours in the middle of the night has satisfied the LD requirement for herbaceous perennials tested. Incandescent lighting tends to cause more stretch than other light sources.

From a presentation at a Professional Plant Growers Association meeting, San Jose, California. January 1996.

Perennials: Surviving Freezing Temperatures

Researchers at Iowa State University subjected five perennial species to the ultimate test of cold hardiness, testing their ability to regrow after exposure to low temperatures. They froze fifteen of each plant at each temperature. Tests revealed that most species survived when exposed to 14° F (-10° C), but damage increased as temperatures decreased. Physostegia was the most hardy, tanacetum the least. Researchers advise growers producing perennials in containers to provide winter protection to prevent root media temperatures from falling below 14° F (-10° C).

Effect of low temperatures on percentage survival of perennials

Herbaceous perennial	Treatment temperature (°F/°C)									
	32/0	28/-2	25/-4	21-6	18/-8	14/-10	10/-12	7/-14	3/-16	0/-18
	Percentage of test plants surviving (%)									
Goblin *Gaillardia* x *grandiflora*	78	100	100	88	100	100	78	88	78	33
Summer Snow *Physostegia virginiana*	100	100	100	100	100	100	100	100	100	11
Stratford Blue *Salvia* x *superba*	100	100	88	88	100	100	67	78	67	11
Robinson's Mix *Tanacetum coccineum*	100	100	100	100	88	88	44	0	0	0
Veronica repens	100	100	100	100	100	100	78	22	22	0

Culture Notes, January 1997.

Hosta Culture Tips

Jim Nau

With the large number of hosta varieties available, it's difficult to pinpoint the best techniques for growing quality crops. Here are some tips for growing and finishing this top perennial shade performer.

Bare-root plants are readily available for either fall or winter shipping, depending on the variety. These varieties can be a species or represent hybrid crosses or sports (mutations), and crop times will differ. Therefore, it's difficult to give a general crop time for all hosta cultivars.

However, if potted in a 1-gal. container, most varieties will be salable in seven to nine weeks for roots potted up in late February and March and grown at 55° to 58° F

(13° to 14° C) nights, with days warmer by 8° to 15° F. For larger plants as well as the slower growing varieties used as borders or edging, allow eight to eleven weeks in a 3- to 4-qt. pot and the same night temperatures.

Tissue-cultured plants are often sold in liners or trays with seventy-two plants per flat. If potted up into 1-qt. pots, they'll finish off in five to eight weeks depending on the variety. Grow on at 55° to 58° F (13° to 14° C) nights.

Finishing

If your goal is to have a plant with limited leaf number (usually less than eight) with a "just rooted" rootball, then these crop times and temperatures will suit your needs. If you're looking for a well-rooted plant with full, lush foliage, you'll need to add two more weeks to the crop time above and four more weeks for more compact varieties. Or better yet, pot up your divisions or transplants in the autumn and overwinter dormant for spring sales.

Crop times are based on a quick turnover rate for mass merchandising. If you can afford the time and labor, give plants some additional time to improve their appearance.

From Ball Perennial Manual *by Jim Nau. Jim is trials and new varieties manager, Ball Horticultural Co., West Chicago, Illinois. November 1996.*

Australian Perennial Culture Tips

Roger Elliot

Perennial production can be tricky, and their exploding popularity makes quality product essential for your customers. Try these tips for three top perennials.

Helichrysum

Helichrysum grows best in full sun with good drainage. It may need 30% shade in Sunbelt areas during summer. Overwatering creates problems, so keep soil barely moist. Use an acidic mix with a 5.5 to 6.5 pH. Pinch at potting or within four weeks after potting. Prune no later than six to eight weeks before finish to avoid reducing the number of flowers. Plants are ready for sale twelve to sixteen weeks after plugs or liners are potted.

Maintaining very compact growth, Nullarbor Gold 'n' Bronze helichrysum has foliage that rarely gets taller than 1 ft. Plants can spread to 3 ft. across after two years. Plants flower throughout the year, peaking during spring, summer, and early fall. Flower heads are 2 to 3 in. across, held on slender stems above narrow, bright green leaves. Buds are pale to mid-bronze, opening to reveal numerous bright golden yellow, papery, petal-like bracts and a golden disc. Use it in baskets and 4-in., 6-in. or 1-gal. pots or for cut flowers.

Brachycome

Brachycome usually grows best in soils that don't dry out too readily, but it dislikes waterlogging. It prefers sun or partial shade and may need 30% shade in the Sunbelt during summer. Brachycome likes soils with 5.5 to 6.5 pH in a free-draining, well-aerated media. Plants grow well in areas that don't have heavy frosts, though they can handle temperatures down to 25° F (-4° C) without too much damage. Liners or plugs should have at least four main breaks at potting, and after another four weeks, shear plants to about 2 in. above media.

Scaevola

Scaevola prefers sun or partial shade. Grow these plants under cover if frost is likely, but compact plants are more likely if grown outdoors. Keep night temperatures from 35° to 75° F (2° to 24° C) inside to promote compactness. Plants may require 30% shade in the Sunbelt during summer. Light applications of low phosphorus fertilizers are beneficial, but be careful not to force-feed plants, because this promotes foliage at the expense of flowers. Good liners should have three to four main breaks at potting. Pinch back to 1 in. above previous breaks at potting or within four weeks of potting. You can pinch or shear lightly anytime except within eight to ten weeks of finish.

Outback Royal Fan scaevola has 1¼ in. wide, bright bluish-purple flowers with a pale yellow basal blotch on the ends of branches. Flowering peaks from late spring to fall. Unlike most scaevolas, it has an upright, fairly compact growth habit, reaching about 18 in. and 2 to 3 ft. wide in two to three years. Prime selling time is June to October, and production time is usually twelve to sixteen weeks. Use in patio pots or hanging baskets.

Dwarf and compact with finely lobed bright green leaves, scaevola Billabong Bright Eyes has ½-in., mauve-blue flowers. Plants in 6-in. pots are salable eight to twelve weeks after potting. It can sucker lightly as it develops to 6 in. tall and may spread to 3 ft. across. Use in mass plantings, 4-in., 6-in. or 1-gal. pots or baskets. The main selling window is from March to November, but some nurseries find they can extend their selling window.

Roger Elliot, Koala Blooms, Victoria, Australia. June 1996.

Petunia

Interspecific Petunia Hybrids

Ann Turner Whitman

Beer isn't the only product coming from breweries these days. Bioengineering research at Japanese breweries has resulted in not only a better brew, but improved petunias. The new hybrids can grow up to an inch a day, cascade or spread 3 to 4 feet and bloom all season long in temperatures from 27° F (-3° C) to over 100° F (38° C). Unlike other petunias, which are cultivars of a single species, these super-

plants are genetically engineered blends of more than one species called interspecific hybrids. Spectacular in hanging baskets, boxes, and mass plantings, Supertunia, Purple Wave, and Cascadia petunias are taking the country by storm.

Propagation

Purple Wave comes true from seed, unlike Supertunias and Cascadias, which must be vegetatively propagated. At 75° F (24° C), Purple Wave gives 90% or better germination when covered lightly with coarse vermiculite. Cuttings of the other hybrids root best at 68° to 74° F (20° to 23° C) air temperatures and 75° F (24° C) soil temperature. Plant liners no deeper than the liner soil level to prevent crown rot. Mist or fog for the first five to seven days and keep evenly moist. Use fast-draining media and drench with a broad-spectrum fungicide at liner planting to prevent disease.

Growing On

Light and temperature requirements are similar to those of other petunias. Grow in full sun at 5000 to 10,000 f.c. Day length sensitive, petunias need long days of at least thirteen hours to flower at temperatures above 65° F (18° C). Suppliers recommend day temperatures of 75° to 85° F (24° to 29° C) and nights at 55° to 60° F (13° to 16° C) for Supertunias, and days at 65° to 72° F (18° to 22° C) and nights at 60° to 62° F (16° to 17° C) for Cascadias.

Growers report that these hybrids are among the most nitrogen-hungry crops that you'll ever grow. To support their rapid growth, a constant feed of at least 300 ppm of nitrogen, 100 to 150 ppm of phosphorus, and 200 to 250 ppm of potassium is necessary. In addition, either give periodic feeds of 400 to 450 ppm N, 200 to 250 P, and 300 to 350 K or top dress with one tablespoon of 19-6-12 Osmocote per 6-in. to 8-in. container. Supplemental iron at the rate of 15 ppm may be necessary if the leaves show chlorosis. In containers or landscape plantings, use double the amount of slow-release fertilizer recommended on the label per cubic foot of soil.

To prevent root diseases, use a well-drained peat-perlite mix with a pH of 5.8 to 6.2 and avoid overwatering. Water thoroughly, but let the media dry slightly between watering without allowing the plants to wilt. These hybrids prefer to be kept on the dry side. If chlorosis shows up, suspect iron deficiency and high pH. Interspecific petunia hybrids are subject to the same pests and diseases as other petunia varieties. Insects to watch for are whiteflies, leafminers, aphids, thrips and fungus gnats.

Control the vigorous growth of these petunias and promote branching by pinching two weeks after planting liners or applying a growth regulator. All growth regulators are effective on interspecific petunia hybrids, but individual varieties may respond differently. Follow your supplier's recommendations for best results.

One cutting per 4- to 6-in. pot will finish in five to six weeks while two to three cuttings per 8-in. basket or four to five cuttings per 10-in. basket will finish in six to eight weeks. Under long-day conditions, Purple Wave flowers sixty days from sowing when grown at 65° F (18° C).

Varieties and Sources

Colors of interspecific petunias hybrids range from white through pink and blue into purple. Supertunias were introduced by Proven Winners and are available from Four Star Greenhouse in Carleton, Michigan; Pleasant View Gardens in Pittsfield, New Hampshire; and EuroAmerican Propagators in Encinitas, California. Varieties include Kilkeny Bells, Purple Sunspot, Pampas Fire, Victory, Velvet Columbine, Pink Victory, and Sun Snow.

Cascadias petunias come in seven varieties including Casablanca, Chaplin, Chamonix, Charme, Chateau, Cherie, and Chico. Purple Wave is a single variety that comes true from seed. Both are available through most U.S. distributors.

Ann Turner Whitman, a horticultural writer from Bolton, Vermont, is a contributing writer for GrowerTalks. *January 1995.*

Koranski on Growing Fantasy Petunia Plugs

Dr. Dave Koranski

The milliflora class of Fantasy petunias from Goldsmith has some excellent qualities that differentiate them from multifloras and grandifloras. Plants are very compact and free-flowering. Flowers are smaller than multifloras (1½ in. diameter) but are earlier and bloom continuously throughout summer with good heat tolerance. They make an excellent display in containers, baskets, porch pots and hanging pouches.

With Fantasy, as with other new crops or varieties, you may need to change some environmental and cultural practices to get enough shoot and root growth in the plug stage. Generally, vegetative growth can be promoted by using more ammonium-nitrogen feed instead of nitrate-nitrogen and more phosphorus. Photoperiod will also control flower induction and initiation and can sometimes interact with temperature to promote or delay flowering. Container size will influence shoot-to-root ratio. When this ratio is high (favoring shoot growth), plants tend to stay vegetative. Many times when plants are stressed  with low moisture, more nitrate-nitrogen, high light, temperature and small container size, they'll bloom faster.

Experiments to determine Fantasies' crop timing reveal they can be grown in any plug tray size, but to establish an optimum root system you need a larger plant. A 288 tray produces a better plant than a 512 tray because there is less stress on plugs. Roots should be active, and the root ball should be developed sufficiently before transplanting. *Don't transplant without a good root system!* This plant needs good root development to withstand stress.

In another experiment, plants fed with 20-10-20 at 100-ppm nitrogen for three to four weeks in the plug tray produced the best quality. Root and shoot growth were increased approximately 40 to 50%, compared to using only 13-2-13-6-3. Also, B-Nine wasn't necessary in our experiments.

Cultural Guidelines for Growing Quality Fantasy Petunia Plugs

Water quality
Maintain low sodium and chlorides (less than 20 ppm).

Media
Starter plug media should be well-drained and contain at least 20% perlite. Soluble salts should be on low side (0.5 to 0.7 mmhos, paste extract), and pH should be 5.8 to 6.0. Starter charge should have enough phosphorus (8 to 12 ppm). Maintain calcium levels of 80 to 100 ppm. During Stages 2 to 4, plugs can be fed with 1.0 to 1.2 EC.

Watering
During Stages 2 to 4, allow media to dry down before watering.

Nutrition
Fertilize with 20-10-20 at 100 to 150 ppm nitrogen during Stages 3 and 4. Supplement with 15-0-15 at 100 to 150 ppm nitrogen for one to two feedings to maintain calcium levels.

Temperature
Keep soil temperature during Stages 3 and 4 at 70° F (21° C) day/62° to 65° F (17° to 18° C) night. Warmer days than nights (positive DIF) will produce more vegetative growth. Avoid using negative DIF before mid-Stage 3, especially with small roots and shoots.

Lighting
Where supplemental HID lighting is used, early flowering may occur under long daylengths, resulting in a less than acceptable shoot to root ratio. Keep HID lighting to eight to ten hours per day until plant is transplantable.

Dr. Dave Koranski, president, ETA Inc., Woodbury, Minnesota. March 1996.

Philodendron

Practically Perfect Philodendron

Lynn P. Griffith Jr.

About two hundred species of philodendron are found, mostly from tropical America. Three principle types of philodendrons exist. The vining or scandent types, such as *Philodendron scandens oxycardium*, the heart-leaf philodendron, are commonly grown as ornamental vines. The second group includes self-heading philodendrons. These plants, such as *P. wendlandii* and the hybrid Black Cardinal,

grow upright on their own. The third class is made up of the erect-arborescent philodendrons. Plants in this group, best represented by *P. selloum*, appear self-heading when young, but as they mature, they become more woody and treelike.

Philodendrons are efficient users of moisture, and many have aerial roots to absorb water from humid air. Flooding can occur in their native habitats, so some philodendrons are ethylene sensitive.

Uses

Because of the diversity of types, philodendrons are grown for a broad array of uses. Container sizes can range from 2 to 17 inches. They're grown from single or multiple cuttings in freestanding pots, while vining types are also frequently produced in hanging baskets or on totems. Philodendrons are useful as ground covers for interior and exterior landscapes. *P. selloum*, also called the split-leaf or lacey-tree philodendron, is popular in USDA Zones 9 and 10 landscapes as a mass planting or a shrubbery border.

Propagation

Vining philodendrons are usually propagated from stem cuttings or leaf-and-eye cuttings. The heart-leaf philodendron is usually grown with multiple leaf-and-eye cuttings per pot or hanging basket, similar to pothos. The cuttings don't require rooting hormones and need very little in the way of mist. Tent propagation is also practiced. The larger vining types are also produced as air-layers, using sphagnum and aluminum foil. Most philodendrons root easily as long as they're given reasonable temperature and light levels. They also root easily in water.

P. selloum is generally grown from seeds, which are collected from tropical landscapes. Frequently, six to eight or more seeds are planted per cell in flats, with the seed barely covered with perlite. A minimum of 300 to 600 f.c. of light is required for germination. Most of the hybrid philodendrons are propagated from tissue culture and sold as liners.

Culture

Heart-leaf philodendrons should be grown between 1,500 and 3,000 f.c., with 2,000 f.c. being ideal. Maintain stock plants under somewhat brighter light. The minimum soil temperature should be 65° F (18° C), with a minimum air temperature of 75° F (24° C). It's best to avoid temperatures below 50° F (10° C), but plants can tolerate 105° F (41° C).

P. selloum is usually grown much brighter, at 3,000 to 6,000 f.c. Many growers like to pot the liners into 6- or 10-in. containers and grow them in full sun until they reach maturity. Plants are then placed under shade for finishing, unless they'll be sold for landscape use. *P. selloum* is also quite tolerant of cold. Depending on its age, it can tolerate temperatures close to freezing without injury, even surviving 22° F (-6° C) as a mature specimen.

Hybrid philodendrons are generally produced at between 1,000 and 2,000 f.c., though there are exceptions. Ideal daytime temperatures are 80° to 85° F (27° to 29° C), with night temperatures from 65° to 70° F (18° to 21° C). Grow hybrid philodendrons on the dry side.

Philodendrons in general like a potting mix with a high moisture-holding capacity and good aeration. Mixtures of sphagnum or sedge peat combined with bark, wood chips, perlite or vermiculite are common. The target pH is 5.5 to 6.0. Allow philodendrons to dry somewhat before irrigating.

Nutrition

Most philodendrons are fairly heavy feeders, requiring generous amounts of dry or liquid fertilizers. A 3-1-2 ratio of $N-P_2O_5-K_2O$ is generally used, as is extra magnesium. Smaller containers and baskets are often grown with 200 ppm nitrogen in constant liquid feed, derived from 24-8-16, liquid 9-3-6, or sometimes soluble 20-10-20. Leaches with clear water every few weeks are desirable. Larger philodendrons may be grown either with liquid feed or with granular, or coated, slow-release fertilizers. Larger philodendrons grow very well on a combination of dry and liquid fertilizer. Leaf size and growth rate are good indicators of fertility status in philodendron. Leaf size and internode length in most varieties quickly decrease if feed rates drop below optimum levels.

Philodendrons usually don't have much in the way of trace element problems. Many show marginal chlorosis of the older foliage when deficient in magnesium. This especially happens in older plants or plants that are cut back several times. It helps to keep magnesium in the spray program on a regular basis. Lack of calcium causes root tips to die, followed by chlorosis, leaf distortion, and shoot tip death. Sprays of a combination of calcium and magnesium chelate result in good, strong philodendrons with large leaves. Iron deficiency symptoms in hybrid philodendrons usually indicate root disease.

Lynn P. Griffith Jr. is president, A&L Southern Agricultural Laboratories, Pompano Beach, Florida, and author of Tropical Foliage Plants: A Grower's Guide. *April 1998.*

Poinsettia

Poinsettia Stock Plants from A to Z

Jack Williams

Healthy cuttings are essential to producing quality blooming poinsettias. Producing cuttings requires attention to the details of planning, monitoring, and adjusting the growth of stock plants in preparation for cutting harvest during the summer.

Establishing Plants

Always start with fresh stock plants. Growers purchasing rooted stock cuttings save four to five weeks of production time over the use of unrooted cuttings. If you're starting with unrooted cuttings in the spring, take extra care in propagation due to

cold, dark weather conditions. Buy cuttings only from reliable propagators who will ship the cuttings properly in cold weather, and unpack cuttings immediately.

Media and air temperatures are critical for successful rooting. Provide bottom heat to maintain media temperature of 70° to 75° F (21° to 24° C). Maintain air temperatures between 68° and 70° F (20° and 21° C) at night and 75° to 80° F (24° to 27° C) during the daytime. Use only enough mist to prevent dehydration. Frequent misting lowers the temperature of the rooting media and slows root development. Excess mist also favors development of numerous diseases and leaches nutrients from cuttings. Transplant cuttings into growing containers once they've developed roots capable of supporting the water needs of the plants.

The container you select—bags, pots, raised benches—depends on your spacing requirements, water delivery systems, crop cycle time, and growing media. If stock plants will be re-spaced during production, select pots durable enough to withstand frequent handling. If you use capillary mats or ebb-and-flood benches, select pots that are short enough to facilitate proper uptake of fertilizer and water. The containers should provide adequate moisture-holding capacity for the plants throughout the summer. Use a well-drained, soilless media free of insects and diseases with adequate porosity and balanced chemical properties.

Light

Night lighting for photoperiod control is necessary prior to May when temperatures in the greenhouse are below 60° F (16° C) or if days are unusually cloudy. Lighting prevents flower bud induction and sustains vegetative growth.

Light intensity for stock plants varies with the stage of production, cultivars, and greenhouse temperatures, but light levels between 2,500 and 3,500 f.c. are usually best.

Fertilization

Begin fertilization at planting using a balanced and complete fertilizer compatible with your water quality. Maintain soilless media pH between 5.8 and 6.3 with electrical conductivity (EC) or soluble salts level below 3.0 mmhos/cm (saturated paste extract).

Temperature

Maintain temperatures while stock is establishing itself as outlined above. Growers risk delays or reduction in cutting production and plant vigor if proper temperatures are not maintained (66° F/19° C night temperature and 80° F/27° C day temperature).

Pinch stock plants to increase cutting numbers and to properly time the age of cuttings for harvest. All stock plants will require an initial pinch to remove the growing tip to initiate shoot development. Pinch when stock cuttings establish roots and do not show signs of stress or wilting. The number of nodes left after the *initial pinch* determines the number of side shoots that result. Five or six weeks after the initial pinch, pinch again to increase the number of shoots per plant.

Growing Stock Plants

Fertilization

Fertilization requirements begin to change with onset of warm weather and more frequent irrigation. Once the crop is actively growing, maintain fertilizer at 200 ppm N constant liquid feed. Stock plants require supplemental calcium and magnesium because tissue is continually removed through pinching. Avoid high levels of ammonium nitrate, which produces large, soft leaf tissue that crowds propagation space and is more susceptible to disease.

Light

Once established, increase light intensity for improved growth and cutting quality while keeping greenhouse temperatures within an acceptable range. Grow dark-leaved cultivars at levels below 4,500 f.c. and all others at levels up to 5,500 f.c.

Increase airflow to keep leaf and air temperatures within an acceptable range. During the final stages of production, provide adequate light to all shoots to improve cutting uniformity. Remove excess foliage to enhance growth, increase cutting caliper, and reduce leaf expansion.

Temperature

Keep temperatures at night above 64° F (18° C). Day temperatures will fluctuate with prevailing conditions and available light. Avoid temperatures above 90° F (32° C).

Growth regulators

Do not use Cycocel alone as it can result in foliar damage and botrytis in propagation. B-Nine (Alar) is preferred to control stretch on stock plants. Combinations of B-Nine and Cycocel are acceptable at low rates as long as the applications are not repeated on a regular basis. Time growth regulator applications about two and a half to three weeks before cutting harvest. Young leaves that form after growth regulator applications are tougher, smaller and perform well in propagation. Used properly, there is no evidence of excessive growth-retardant carryover in the cuttings.

Insects and diseases

Excess use of pesticides can result in foliage burn, which may promote botrytis development in propagation. To minimize pest problems in propagation, use a good scouting program and treat stock plants as needed from the start.

Final pinch

The final pinch before harvest is referred to as the *critical pinch* and determines the maturity of cuttings to be harvested. For best success in propagation, it is important not to use cuttings that are too young (immature) or cuttings that are too old. These cuttings do not initiate roots uniformly and are difficult to handle in propagation. Stock plants started in March or April should yield at least twenty cuttings per plant. If the grower is able to use available cuttings earlier and later in the season as well, this number can easily be doubled.

Keep in mind that cutting harvest takes place during the hottest time of the year. Cutting first thing in the morning is best as plants are turgid and may easily be removed by hand or with a knife. As temperatures rise, the plants become rubbery and more difficult to cut. If this happens, it is best to use sharp knives or wait until the next morning when plants are once again turgid.

Producing healthy poinsettia cuttings is a challenge for stock growers everywhere. Through commitment to excellence, hard work and adaptability, it is possible to produce quality cuttings that endure the stress of propagation and bloom in the care of our industry professionals.

Time required from critical pinch to harvest

5 weeks	6 weeks				7 weeks
ECKESPOINT	ECKESPOINT	GROSS		MIKKEL	ECKESPOINT
Lilo Red	Freedom Red	Supjibi Red		Blitzen	Jingle Bells 3
Lilo Pink	Freedom White	Supjibi Pink		Dasher	Lemon Drop
Lilo Marble	Freedom Pink			Donner	
Lilo White	Freedom Marble	ANNETTE HEGG		Yuletide	
	Freedom Jingle Bells	Dark Red		Yuletide Pink	
PEACE	Celebrate 2	Topwhite		Yuletide White	
Jolly Red	Celebrate 2 Pink	Hot Pink			
Red Elegance	Celebrate 2 White			PELFI STARS	
Red Splendor	Monet	GUTBIER V-14		Bonita	
	Pink Peppermint	Glory		Cortez	
	Red Sails	White		Dark Puebla	
		Pink		Flirt	
	GUTBIER V-17			Maren	
	Angelika Red	PEACE		Nobelstar	
	Angelika White	Cheers		Picacho	
	Angelika Pink	Frost		Puebla	
	Angelika Marble			Sonora	

Jack Williams is technical advisor for the Paul Ecke Ranch and Poinsettia Growers Association, Encinitas, California. May 1995.

Proper Nutrition for Better Poinsettias

P. Allen Hammer

New poinsettia cultivars, bract necrosis, and a better understanding of fertilizer requirements have all greatly changed recommended fertilizer practices for poinsettia production. Constant liquid fertilizer application rates of 200 ppm nitrogen and potassium (plus phosphorus) or less are recommended for newer, dark leaf cultivars. Levels above 200 ppm can result in overfertilization that can increase root loss, black stems, leaf yellowing, and stem brittleness, resulting in breakage during handling and shipping.

It's important to apply adequate calcium in your fertilizer program, maintaining a 2:1 calcium:magnesium ratio to avoid competition between calcium and magnesium. At the same time, you should apply adequate magnesium for production. Apply calcium at 100 ppm or greater, with half the amount of magnesium at each watering. Although it's very important to supply adequate calcium in your liquid fertilizer program, weekly sprays of 400 ppm calcium from calcium chloride at the beginning of bract coloration are recommended as a calcium supplement to reduce bract necrosis. Avoid excess potassium and sodium levels because they compete with calcium uptake.

Test your water source for excess elements and pay special attention to fluoride (less than 5 ppm) and lithium (less than 2 ppm). Both elements have been associated with leaf edge burn and bract necrosis in poinsettias.

Poinsettias have high molybdenum requirements. Apply molybdenum at the rate of 0.1 ppm in a constant feed program. You can use higher rates on a periodic basis. Ammonium toxicity is also a real problem in poinsettia production, and no more than one-third of the total nitrogen should be applied as ammonium or urea. Two-thirds of the nitrogen should be in the nitrate form.

Postharvest poinsettia performance will improve if you reduce fertilizer application rates and make sure root medium salt levels are minimal at flowering and shipping. Many growers are reducing fertilizer rates to 100 ppm nitrogen and potassium or less at bract coloration and leaching salts from the root medium before shipping.

Root medium analysis is also a very important part of any poinsettia fertilizer nutrition program. I strongly suggest sending monthly root medium samples to a competent laboratory for analysis. Results from these analyses allow adjustments in fertilizer application to maximize growth and minimize nutrition problems.

P. Allen Hammer, professor of floriculture, Purdue University, West Lafayette, Indiana. October 1995.

Scout Early for Powdery Mildew on Poinsettias

Joli A. Shaw

Scouting for powdery mildew on poinsettias is critical for effectively controlling the disease. Because powdery mildew can "explode" once it's established even at a low level, the key to control is early detection, Mary Hausbeck, assistant professor of botany and plant pathology, Michigan State University, said at the SAF Pest Management Conference in February of 1995. Growers often don't find powdery mildew problems until late in production when bracts become infected and losses are significant. If the disease takes off, fungicides may have limited effectiveness or may leave unacceptable residues.

Before potting, scout one of every ten cuttings. After potting, scout one of every thirty plants weekly in areas where powdery mildew hasn't been detected. Symptoms include white, talcum-like mildew colonies up to approximately ½ in. in diameter on leaf or bract surfaces. Scout once a week, examining four fully expanded leaves in each of the mid and lower sections of the plant. Check both top and bottom leaf surfaces. Powdery mildew begins on lower leaf surfaces and isn't visible on upper surfaces as chlorosis until the colony below has caused significant damage.

When scouting, note whether colonies are living and active, as colonies that aren't active will be flat against the plant's surface, while colonies that are alive and active will have a powdery or fluffy appearance because of spore production, Mary Hausbeck said. Signs of powdery mildew may also develop on infected but symptomless poinsettias after they leave the production greenhouse, causing customer dissatisfaction. After you detect powdery mildew, scout one of every ten plants until they're disease-free for a minimum of three weeks. After that time, resume scouting one out of thirty plants.

If you find powdery mildew, remove all infected leaves and place them in a sealed bag immediately. Don't carry infected leaves with active colonies through the greenhouse for disposal, as spores on infected tissues will be released, possibly infecting nearby healthy plants, Mary stated. With regular scouting, if powdery mildew occurs, the number of infected leaves you find should be low enough to make leaf removal feasible (one to five leaves per infected plant). Removing leaves at an early stage of disease development is critical, as infected leaves could contribute to the spread and severity of the disease. After removing infected leaves, apply fungicides immediately. Spray all poinsettias in the greenhouse where you first find powdery mildew. Follow labels for application intervals, and continue applications until you find no active colonies when scouting.

Joli A. Shaw is a freelance writer and former associate editor of GrowerTalks *magazine, Lisle, Illinois. October 1995.*

Late Growth Regulator Application Tips

Sally Roberts Moore

Poinsettia growers, faced with a late-stretching crop, have traditionally had to sacrifice bract size to gain height control when applying chemical growth regulators on or after October 15, says Jack Williams, Paul Ecke Ranch, Encinitas, California.

Now growers can avoid that trade-off, Jack says, by turning to A-Rest and Bonzi for late growth regulator applications. Here are a few tips to help growers successfully make those late A-Rest and Bonzi applications.

Both chemicals must be applied as a drench, because spray applications will negatively impact both bract size and crop timing. When drenching, correct dosages are important. Bonzi controls height when applied at 25 ppm, Jack says. A-Rest offers height control at ½ to 3 ppm. He recommends a 1 ppm A-Rest drench as the starting point for most growers. However, Southern growers may find 2 ppm most effective.

Cultivar differences are evident with these chemicals. Red Sails and the Hegg varieties seem to be the most sensitive to most chemicals, Jack notes.

In trials with these chemicals, researchers applied them at the end of October about four weeks after the start of short days. Because an additional inch or two of growth occurs after applications, Jack recommends that growers apply chemicals before the plants reach a maximum desired height. Applications made later than the end of October will give less additional growth.

Be careful when mixing and applying chemicals, Jack says. A dosage of 2 ppm of A-Rest equals about 1 oz. of that chemical in 1 gal. of water, applied at a rate of 4 oz. to a 6- or 6½-in. pot. If the dosage exceeds 4 oz., you'll get more height control than expected. Inadequate applications will result in less control than needed.

Adjusting Applications

Larger pot sizes require larger doses. Dosage is based on media volume in the pot. Increase dosages if you're using a media with pine bark. Generally, you should increase rates by 25%, Jack says.

Because you can't spray the chemicals or run them through an injector system, Jack suggests using a sprayer and dispensing the appropriate amount directly to each pot through the sprayer wand. To assure that you apply accurate dosages, Jack recommends having the person using the sprayer practice until he or she is comfortable with how long it takes to release the 4-oz. application.

A uniformly moist soil is also important for good results. In dry media, the application isn't distributed evenly and can't travel all the way to the root zone, Jack says. If the chemical can't reach the roots, it isn't fully absorbed, diminishing control.

When applying the chemicals to small pots, such as those used for growing miniatures, Jack suggests diluting the chemical further and then applying twice the volume of water. You can make a second application about three weeks after the first application if you need to control additional stretch, Jack says. Most growers will get adequate control from one application.

Sally Roberts Moore, freelance writer, Ann Arbor, Michigan. Currently she's editor of Florist *magazine.*
October 1996.

Winning the Whitefly Battle

Jan Hall

By October you should have pinched your poinsettia crop, and side shoots should be well established. At this point, focus your attention on whitefly control. The addition of Marathon to our pesticide arsenal has certainly made it easier to control whiteflies. If whiteflies are still present, however, this is the time to use all your efforts for control before bract development.

- Learn to recognize the different types of whiteflies and their stages on poinsettias.

- Scout the crop at least once a week. Check plants randomly through the greenhouse, and inspect leaves from the top, middle, and bottom canopy of each plant selected.

- Flag a few infested plants to use as indicator plants, and reexamine these plants weekly. This will allow you to monitor the development of the whiteflies and the effectiveness of your control measures.

- If you find high levels of immature whitefly, consider removing the infested lower leaves as a control strategy. Be sure to destroy the foliage you remove to prevent whitefly from moving back onto your crop.

- The list of available chemicals is wide open at this stage of the crop. After bracts develop, you'll lose the ability to apply some insecticides due to residue or the potential for phytotoxicity on bracts.

- Apply insecticides based on the stages of whiteflies present in your crop. For example, don't use a pyrethroid if most of your whitefly population is in the egg or pupal stages.

- Apply Marathon in October if you plan to use this pesticide and have not yet done so. As poinsettia stems become woody, it will take longer for Marathon to move up into the leaves and control whiteflies. Don't expect Marathon to solve all of your whitefly problems. You may need to use other insecticides before Marathon applications or later in the crop.

Jan Hall, technical services, Paul Ecke Ranch, Encinitas, California. October 1996.

Calcium Sprays Eliminate Bract Edge Burn

Debbie Hamrick

Many poinsettia growers never even know the bracts on their poinsettias are calcium deficient. The problem often doesn't show up until after plants have been shipped and the customer is unpacking them. To avoid complaints about bract edge burn this year, make a note on your poinsettia production schedule to begin 400 ppm calcium sprays as soon as the first transition bracts begin to show color. Simultaneously, discontinue your regular feeding program. Plants should have received enough fertilizer by then to carry them through finishing, retail, and in the home. Continue calcium sprays weekly (four or five applications) until you ship.

Bract edge burn is a function of how much calcium is in the bract. Bracts can be calcium deficient even though you may have supplied plenty of calcium in growing media, and leaves aren't calcium deficient. That's because calcium moves inside the plant's transpiration stream. Because bracts have no stomata, calcium isn't transported there. Humid greenhouses will cause more difficulties than low humidity greenhouses.

Discontinuing fertilization when beginning calcium sprays is important because fertilizer competes with calcium uptake. At the end of the crop, the goal is to lower soluble salts and increase pH of growing media. Calcium also protects plants from botrytis during shipping.

How sensitive are the poinsettia varieties you grow?

Bract edge burn sensitivity varies by variety, according to Jim Barrett, University of Florida, Gainesville. Here's Jim's experience with variety susceptibility:

Very high sensitivity	V-14 group
	Supjibi (especially in cool climates)
	V-17
	Celebrate II
High sensitivity	Freedom
	Success
	Fischer varieties
Low sensitivity	Hegg group
	Jolly Red
	Petoy

Debbie Hamrick is editor of FloraCulture International, *Batavia, Illinois. October 1996.*

Shipping Healthy Poinsettias

Though growers used to see bract edge burn (botrytis) mainly in the greenhouse just before shipping, the problem has in recent years moved to the next level: Retailers and wholesalers are more likely to see the problem after the shipment arrives at their warehouse or store. Terril Nell and Jim Barrett, University of Florida, Gainesville, have been investigating strategies for preventing bract edge burn on your poinsettias during shipping. Here are their suggestions, as reported in the *Ohio Florists' Association Bulletin.*

Knowing that you can't examine plant quality in the greenhouse is the first step to preventing botrytis, according to Terril and Jim. Because studies have found that botrytis increases after shipping, you must evaluate your plants as your customer will see them.

Research has revealed that spraying calcium is one of the most effective ways of reducing or eliminating bract edge burn problems. Calcium deficiency occurs at bract edges where water transpiration is not adequate, meaning calcium isn't carried to the entire bract.

Unfortunately, typical greenhouse conditions in late November and early December—cool temperatures and high humidity (above 90%)—are the perfect conditions for botrytis development. In addition, shipping plants in boxes can also contribute to bract edge burn.

So what's the best way for you to keep bract edge burn off of your crop? Terril and Jim have grouped varieties into sensitivity levels for the likelihood of botrytis development. Start by making sure you apply 400 ppm calcium sprays weekly starting at first color to any variety in level two, three or four (see chart). Use calcium nitrate, chelated calcium or calcium chloride.

If you're growing plants with level two, three or four sensitivity and you'll be boxing them for shipping, take steps to reduce *Botrytis inoculum* in your greenhouse:

Dehumidify at night and during overcast days. This may also mean increasing fungicide applications, as some trials have shown that fungicide use is as effective as calcium sprays for reducing botrytis development during shipping.

Lower your fertilization, potassium and ammonium levels to reduce the chance of developing calcium deficiency or botrytis. In particular: Lowering your fertilizer levels during November will help. Reduce fertilizer so that plants finish with ¼ or less of the fertilizer they had at the beginning of the crop. One note of caution: Though the newer dark-leaved varieties with large bracts require lower fertilization levels, the standard light green-leaved plants need higher fertility levels.

Try not to box highly sensitive and most sensitive varieties (levels three and four) for shipping. If you do, minimize the time plants spend in boxes.

Lateral stem breakage is another problem during shipping, particularly in the South. It can be remedied by preventing stretching and knowing your varieties' growth habits and weaknesses. Here are Jim and Terril's tips to keep in mind during shipping:

- Freedom, the best variety for shipping, will stretch during November. In fact, in shipping trials, dark-leaved varieties such as Festival, Freedom, Pepride, Sonora and Spotlight held up the best, with less leaf and bract fading and less disease sensitivity. Use them for boxed orders.

- Some of the newer light green-leaved varieties didn't fare so well in shipping trials. They include Bonita, Flirt, Marblestar, Maren, Noblestar, Nutcracker series, Peterstar series, Picacho and Puebla.

- Some varieties stretch at the end of the crop, especially if grown with tight spacing. They include Cortez, Festival, Freedom and Monet.

- To prevent horizontal growth (which breaks off), don't space out Supjibi and Petoy too soon.

- The strongest varieties on the market are Red Splendor, Sonora and Success. On the other hand, Monet and Celebrate 2 can have severe stem breakage. Don't grow them too large or too soft.

Bract edge burn sensitivity

Level 4: Most sensitive Red Delight, Subjibi (in cool climates), V-14 Glory series, V-17 Angelika series
Level 3: Highly sensitive Bonita, Celebrate 2 series, Flirt, Maren, Nutcracker series, Picacho, Peterstar series
Level 2: Sensitive Cortez, Freedom series, Sonora, Spotlight, Success, Pepride
Level 1: Least sensitive Dynasty, Hegg types, Jolly Red, Supjibi (in warm climates), Petoy

Culture Notes, November 1997.

Calcium Sprays—Not Just for Bract Edge Burn

Joli A. Shaw

Though talk on calcium sprays has recently focused on their effectiveness in preventing bract edge burn in poinsettias, one Minnesota grower has discovered another benefit of the applications.

Dave Linder, Linder's Greenhouse Inc., St. Paul, sprayed the 6½-in. Nutcracker Red poinsettias at his main location with calcium at 6 oz./50 gal. of water four times—about every ten days starting September 27. At his Lake Elmo facility, he left his 6½-in. Nutcracker Red untouched. When the crop finished and Dave compared the two, he found that plants that had been sprayed had a deeper, truer red color. In contrast, the untouched plants were an orange red, not as pleasing to consumers' eyes.

Joli A. Shaw is a freelance writer and former associate editor of GrowerTalks *magazine, Lisle, Illinois. February 1997.*

Poinsettia Cyathia Split

Growers in some parts of the U.S. during the 1997–98 poinsettia season experienced cyathia split on certain varieties, the industry grapevine has told us. Some growers reported that as much as 80% of their production of one variety exhibits this problem. They saw it most often on straight ups (unpinched plants), though it was seen on pinched plants as well. Some experts say the problem can be traced back to stock plant management.

Splitting often occurs because of plant maturity, which begins back at the stock plant. A variety may be "programmed" to begin splitting at the twentieth node. The cutting might have been taken from nodes six through ten of the stock plant, which means the new plant's first new node will be node eleven. Nine more and the plant begins to split. Research has shown that keeping greenhouse temperatures above 70° F (21° C) during flower initiation can "erase" this node memory. Conversely, cool temperatures below 65° F (18° C) may accentuate the problem.

Culture Notes, February 1998.

Poinsettia Stock Tips

Fischer USA's Karl Trellinger says fertilization, temperature, and pest management are some of the areas that most often challenge poinsettia growers. These tips for producing quality mother stock of the new Fischer varieties will help ensure your success.

Fertilization

The exact program depends on watering practices and adjustments based on soil and water analysis. Use constant feed with 200 to 250 ppm for dark-leafed varieties and 250 to 300 ppm for medium-green varieties depending on light level, plant age, and growth rate.

For softer growth, 20 to 40% of the total nitrogen can be ammonium, using a balanced 20-10-20 or 21-5-20 and adding micronutrients, calcium nitrate, and magnesium sulfate. For harder growth, use 5-5-15 with additional micronutrients.

Soil analysis should be done every one to two weeks. Recommended for the start of the crop: 250 ppm nitrogen, 40 ppm phosphorus, 250 ppm potassium, 150 ppm calcium, 80 ppm magnesium.

A tissue analysis report should read: nitrogen 4.0-6.0%, phosphorous 0.3-0.5%, potassium 2.0-3.5%, sulfur 0.25-0.70%, calcium 1.2-2.0%, magnesium 0.6-1.0%, iron 100-300 ppm, copper 4-25 ppm, boron 30-100 ppm, manganese 100-300 ppm, zinc 40-100 ppm.

pH
Maintain pH at 5.8 to 6.2. This is very important.

Temperature and Branching
Temperatures should be 80° F (27° C) day, 68° F (20° C) night (65° F/18° C minimum). With temperatures higher than 85° F (29° C), keep soil on the moist side. Also, you may need additional shade. Syringe plants with water during the hot hours of the day (three to five times per day) to avoid hardening of stock plants. This will reduce stress of the plants by bringing temperatures down, especially on dark-leafed varieties, and produce even branching.

The way cuttings branch is somewhat determined by whether they were or weren't ideally grown on the stock. Keeping them stress free (avoid low light, high humidity and minimal airflow) in propagation and during and after pinching ensures excellent and even branching.

Light
Provide 2,500 to 3,500 f.c. until two weeks after pinch and 3,500 to 5,000 f.c. following that.

Diseases
Pythium: The best treatment is Subdue at 1 to 0.5 oz. per 100 gal. every four weeks starting right after planting. Use coarse peat, and keep plants moist—avoid drying out, overwatering and high salt levels. If soil gets dry, irrigate with clear water before feeding.

Botrytis: Maintain good air movement and low humidity. Water overhead only in the early morning hours. Spray with Phyton at 1.5 oz. per 10 gal. or with a mixture of Daconil and Zyban or Ornalin or Chipco all at half rate. Use Ornalin only after rooting, as it might delay the rooting process.

Powdery mildew: Maintain good air movement and low humidity. Spray with Phyton (2.0 oz. per 10 gal.), Cleary's 3336 (1 lb./100 gal.) preventative every fourteen days or with Phyton at 3 oz. per 10 gal. when disease is present on bracts (test first, although the manufacturer assures that you won't have phytotoxicity problems) or Strike at 4 to 6 oz. per 100 gal.

Rhizoctonia: Maintain proper salt levels and moisture content of media. Drench with Cleary's 3336 (1 lb./100 gal.) or Terraclor (4 oz./100 gal.) every six weeks.

Insects
Biological control programs against whiteflies and thrips are working fine for many growers; extensive trials and good working relationships with suppliers and universities seem to be critical for their success. With Marathon being used successfully against whitefly, other pests like thrips and mites are on the rise, as chemical spray applications are less frequent than in previous years.

Whiteflies: Soil application of Marathon is successfully keeping whitefly populations in check. The suggested treatments against thrips and mites should take care of whatever whiteflies are still around.

Thrips: As Marathon doesn't work well against thrips, the rising populations have to be kept in check with other chemicals such as combinations of Tame and Orthene, Enstar II and Avid, or Sanmite. The latter combination is also good for mites and whiteflies. Make weekly applications.

Mites: Cyclamen mite, spider mite, and Lewis mite populations are rising. Apply Enstar II and Avid, Sanmite or others weekly.

Fungus gnats, shore flies: Spraying floors weekly with Diazinon (12 oz./100 gal.) greatly reduces insect populations. Treat severe infestations with drenches of Diazinon (12 oz./100 gal.) or Duraguard (25 to 50 oz./100 gal.) three weeks in a row.

Culture Notes, April 1997.

Poinsettias

For multiple-bloom, pinched plants, take cuttings in July and August. For plants that will finish in larger containers (7 and 8 in.), take cuttings in the third week of July; for 5- to 6-in. pots, take cuttings the first week of August. Take cuttings for finished 4-in. pots the third week of August. These cuttings can be directly rooted in the final container. Cuttings for larger finished pots can also be rooted in their final containers, but are best started in smaller propagation containers.

Cuttings root best if media is 72° to 74° F (22° to 23° C). Keep air temperatures below 85° F (29° C) to avoid stressing the cuttings. A mist or fog system is usually required to aid in rooting. Light levels should be 1,500 to 2,500 f.c. until cuttings root. First roots appear in ten to fourteen days, but plants may do better after transplanting if they stay in the propagation media one to two weeks longer.

Keep the propagation area clean. Sterilize tools, and use clean, sterile propagation containers. If possible, disinfect the propagation area before sticking cuttings. Employees should disinfect their hands before handling cuttings.

Culture Notes, July 1997.

Poinsettia Production Tips: Chemical Growth Regulators

Jack Williams

Using growth regulating chemicals to produce poinsettias has become routine for growers all over the world. The most common plant growth regulators (PGRs) used by growers prevent or control internode stretch. Without PGRs, plants are likely to become too tall for sale or too weak to withstand the rigors of transportation. By using good cultural practices in applying PGRs, growers can produce good quality poinsettia crops under the most difficult conditions.

Growth rate and potential internode stretch in poinsettias are directly influenced by environmental conditions such as temperature, relative humidity, and light intensity provided for the crop. Grower-influenced conditions such as irrigation practices, spacing, and crop scheduling also play an important role in preventing unwanted stretch. Understanding how to use these factors together to minimize potential stretch is a challenge to poinsettia growers every year.

When crop growth escalates to the point of needing chemical intervention, growers have several PGR options for poinsettias.

Currently, plant growth regulators most commonly used with poinsettias are A-Rest, B-Nine, Bonzi, Cycocel, and Sumagic. Each has a different activity level and application requirement for poinsettias. To determine which PGR is best to apply, growers should evaluate growth stage to which the application will be made and amount of growth control or activity level desired. Using this information, selecting an appropriate PGR is simple.

Transplant to Pinch

PGR use is generally limited at this developmental stage. Most cuttings have been treated with growth retardants during propagation and are less likely to stretch. As roots begin to establish into the new growing medium, active growth begins. If cuttings already have adequate node count, any new growth is likely to be removed by pinching. Therefore, PGR application isn't warranted, as existing internode spacing won't change. Exceptions include plants grown as non-pinched forms, premature lateral branching before pinch or plants that must be grown to a greater node count before pinch can take place. In all of these situations, spray Cycocel at fairly low concentrations (1,000 ppm) only as needed.

Pinch to Flower Initiation

Growers usually begin using growth retardants when lateral shoots reach ¾ to 1 moderate control, spray foliage with B-Nine and Cycocel combined at 1,250 ppm each or use Cycocel by itself at 1,500 ppm. If you need stronger control, spraying Bonzi at 10 to 25 ppm, depending upon cultivar and level of activity desired, generally works well. Recent pinches make it easy to get uniform coverage and good contact with stems and petioles for proper absorption of Bonzi. Drench applications provide longer lasting and more uniform effects than spray applications. For moderate control, apply Cycocel as a drench. For stronger control, use A-Rest, Bonzi or Sumagic.

Flower Initiation through Mid-October

Application options become more limited during this developmental phase. Spray applications of B-Nine aren't recommended except in extremely hot regions such as south Florida, the gulf states, or the tropics. Using B-Nine at this developmental stage can delay flowering and reduce bract size. A-Rest or Bonzi applications are effective at lower concentrations, but they aren't advisable at this time, as they may slightly reduce bract size at finish. Studies in 1994 by the University of Florida and the Poinsettia Growers Association showed bract size reduction was more significant at this stage of development than at earlier or later application dates. Although Sumagic was not included in these studies, its response is usually similar to these PGRs. Cycocel drenches or sprays are acceptable at this stage under a wide range of conditions.

Mid-October through Early November

Growers should try to have all growth retardant applications completed by this time. However, if weather conditions or unusual growth patterns cause late stretch, it's possible to apply A-Rest at 1 to 3 ppm or Bonzi at 0.5 to 2 ppm without significant bract size reduction or flower delay. Maintain rates at the low end unless you are located in southern extreme-heat regions.

Apply these chemicals as a soil drench using a solution volume proportional to the container's soil volume. For 6- to 6 ½-in. pots, apply 4 oz. of the PGR solution to uniformly moist media. For accuracy, it isn't advisable (or legal by some labels) to apply PGRs through irrigation systems. The most common method of application has been by dipping the chemical out of solution tanks or using chemical spray equipment with nozzle tips removed.

Drench applications are extremely effective with hanging baskets, where it's difficult to spray without drift onto adjacent plants and where growing conditions encourage rapid, stretched poinsettia growth. For 10-in. baskets, 15 oz. of the PGR solution should be applied. Growers using pine-bark-based media should increase rates by about 25%, as this bark reduces these chemicals' activity.

Jack Williams, Paul Ecke Ranch/Poinsettia Growers Association. September 1995.

Finishing the Perfect Poinsettia

Teresa Aimone

By September, you're wondering, "Where has the year gone? It's almost Christmas!" For those of you who have been looking at poinsettia stock plants since March, you may want Christmas to come a whole lot sooner. One good point: Poinsettia sales are starting earlier and earlier every year. It's not unusual for growers to ship the week after Halloween, or even before in a few cases. Hopefully, by September your crop is already booked, scheduled for early delivery, and growing beautifully.

Here are a few last-minute production tips for finishing a quality crop.

Growth Regulators

It's best to have all growth regulator applications done by mid-October through early November. If your poinsettias typically have a problem with late stretching, apply a light drench application of either Bonzi (0.5 to 1 ppm) or A-Rest (1 to 3 ppm). This application should be done when plants are approximately 2 in. below their desired finished height.

Temperature

Maintain temperatures of 67° to 68° F (19° to 20° C) nights and 72° to 74° F (22° to 23° C) days for best bract development. Hold these temperatures until the middle of November, when temperatures can be dropped to 63° to 65° F (17° to 18° C) nights and 67° F (19° C) days.

Fertilization

Your fertilizer program will be most effective if you perform a regular soil analysis on your crop. This will allow you to adjust feed levels and schedules, watering practices and even light levels to produce the best quality poinsettias possible. A soil analysis should preferably be done every two weeks; it's a very inexpensive investment in your crop.

Generally, dark-leafed varieties require less fertilizer than varieties with lighter green foliage. Provide constant feed of 200 to 250 ppm nitrogen for dark-leafed varieties. Those with lighter foliage should receive 250 to 300 ppm nitrogen. Poinsettia breeders suggest that up to 30% of the total initial nitrogen source can be ammonium. Reduce that percentage to 15% by the end of October. Gradually reduce fertilizer levels beginning the middle of November. You should be feeding at 30% of the original nitrogen at the end of the crop.

Humidity

Humidity control becomes important at the end of the crop, especially if the weather turns wet and cloudy. Keep humidity below 70%, and provide good air circulation, particularly if you're growing on close spacing.

Teresa Aimone was a regional specialist Southeast, S&G Seeds, Coppell, Texas. September 1997.

Preventing Common Poinsettia Problems

Terril Nell

Want less stem breakage on your poinsettia crop? Leave fewer laterals after the pinch. Try to leave three to five laterals rather than seven or nine. You'll decrease stem breakage by having fewer, and if you left more laterals, after breakage, you may only end up with three anyway.

Preventing Bract Edge Burn

Bract edge burn attacks all poinsettia varieties and colors with different symptoms in different varieties and climates. Many times botrytis develops along with bract edge burn.

Bract edge burn is usually most severe on the oldest bracts and appears after several cyathia show pollen. In the worst situations, it can occur before bracts are fully expanded. It's also common to see injury on bracts of short laterals covered by leaves and bracts on taller laterals.

It's most likely to occur on Supjibi and V-14 Glory, while the Hegg types and Jolly Red appear to be affected the least. On varieties such as V-14 and Sonora with bracts that have distinct points along the sides and ends, brown dead spots may first appear on these tips. On more rounded bracts, the necrotic areas often develop first along the margins. The most common symptoms on Supjibi are a row of small 1- to 5-mm spots running parallel to and just in from the bract margin.

All poinsettia growers should be evaluating their environmental management and cultural practices to reduce the chance of developing a calcium deficiency problem.

Terril Nell, Department of Environmental Horticulture, University of Florida, Gainesville, Florida. September 1997.

Primula

Primula Culture Made Easy

Tom Linwick

Primula has come a long way in variety, seed quality and technology. Breeding has made primula a shorter crop with improved seed quality. Most seed companies are now working with pregerminated and primed seed to make primula production much easier for the growers. The primulas of today are very uniform flowering and have much larger flowers on sturdy stems. The types of primulas grown vary some-what by geographic region, as in certain areas they're used more as bedding plants, and in other areas they're used as pot plants. In a warmer climate, such as California, most landscapers prefer the veris or polyanthus types, as they have their flowers on a stem that stands above the foliage. They feel these types show up better and hold up longer without having to be cleaned. In the Northwest in places like Washington, Oregon and Canada, growers tend to grow mostly the acaulis types of primula, which have a cluster of flowers that come from the crown of the plant and sit on top of the foliage. These primulas are used as pot plants in the early season and as bedding plants later in the spring. The majority are used as pot plants. Many are also used in early combination planters with pansies, dusty miller, and various bulbs.

Breeding efforts have tried to introduce earlier flowering primula so their crop time isn't so long, but recently most companies have begun breeding for later flow-ering types. By having later flowering primula, growers can sow an early flowering series at the same time as the later flowering type and spread out their finished crops. Many times growers get into a bind when their whole primula crop is ready at about the same time. Growers are now trying to find crops or varieties that flower at times that aren't the same as the majority of other growers so they don't have to compete solely on price.

People are starting to become more interested in some of the old types of prim-ula such as the veris and wanda types, which are hardier than the acaulis types. They can be used as bedding plants in the fall with pansies, and because they tend to be a bit hardier, they can stand up to more adverse weather conditions. This doesn't mean acaulis types have declined in popularity, as many improvements have also been made with these types.

The demand for new primula varieties is still increasing, with most production companies continuing to improve their existing F_1 hybrids and add new colors. The goal is to have uniform flowering throughout the separate colors in a series so growers have an easier time selling their product. This ensures not having too many of one color at the start or finish of the season.

Many companies have introduced improved series to replace their existing ones, while others have continued to add new colors to an already improved series. The Ducat and Crown series from S&G Seeds have been replaced by Corona and Lira, and Goldsmith Seeds has replaced their Sagas with Quantum and introduced a later flowering Gemini series. Daehnfeldt has continued to add new colors to their Danova series and has introduced a late flowering series, Daniella.

A lot of breeding in the veris or polyanthus type of primula has been done, but not many of these new series have been introduced in the U.S. market. Also some of the existing series have been discontinued or the range of colors has been reduced. Daehnfeldt has introduced a series called Concorde that has ten separate colors and is bred to have very large flowers on sturdy stems.

These new F_1 hybrid primulas have many advantages, including increased germination, uniform flowering, and larger, more attractive, longer-lasting flowers. The increased germination in the 80 to 90% range really helps growers producing plugs, while uniform flowering helps those finishing primula. The larger flower size helps both those that are wholesaling and retailing the finished product.

Primulas aren't a very difficult crop, if you pay attention to the requirements for propagation and finishing.

Propagation

Both temperature and humidity are very important for germinating primula. Lighting is beneficial but not necessary in initial primula germination.

Seed flats

The soil used in germinating primula should be a finely sieved peat with vermiculite added for adequate drainage. Most plug mixes work well for this. Sow the seed into a peat lite mix with very low salts (1.5 to 1.9 EC) and a 5.5 pH. This is very important because high salts will inhibit germination. Seeds don't require covering to germinate. If covered, use a thin layer of medium vermiculite so light can pass through to seeds. The vermiculite can help keep the moisture levels up around seed and also anchor newly emerging seedlings. Some growers add a little vermiculite after germination or after seeds are removed from the germination chamber.

Temperature

Germinate at 64° F (18° C) making sure that temperature doesn't exceed 68° F (20° C) because high temperatures will inhibit germination. Germination takes about ten days, but can vary depending on temperature.

Humidity

Humidity is very important for good germination. When germinating in a chamber, try for as close to 100% humidity as possible. Some growers cover their seed trays with plastic to keep humidity levels up. After about ten days they remove plastic and put seed or plug trays into a shaded greenhouse. Shade is required to protect seedlings and ensure complete germination. Pay close attention to humidity, because some seeds may still be emerging. Don't let seed flats dry out.

Lighting

Lighting isn't necessary for good germination, but it can help even at low levels to prevent seedlings from stretching. Simply using florescent lighting for twelve hours is beneficial. If you watch seed flats carefully, you can remove them from the germination chamber or uncover the plastic on them once seeds are just emerging.

Fertilization

At about three weeks or once seeds are all germinated, you can begin feeding the seedlings with 50 to 75 ppm 20-10-20 peat lite mix or with 1.25 EC and 5.5 pH.

Growing On

Transplanting

At approximately five to six weeks from sowing, seedlings should be ready to transplant. In plugs it takes seven to eight weeks from sowing depending on plug size. Larger plugs can be planted a bit later, but take care not to hold plugs too long because this can slow their overall development. Soil should be slightly fertilized with 5.5 to 6.0 pH. Most growers transplant into the final container using 3½-in. to 4-in. pots. Plant deep enough, but don't cover the plant crown. Newly potted primula must never dry out. Shade at about 50 to 60% until well established in their final containers. Shading will also keep temperatures at more acceptable levels.

Temperature

In the young stages, grow plants at approximately 59° F (15° C) until plants are well established. For the best quality primula crop, provide plenty of air circulation, keeping greenhouses or cold frames well ventilated. Once plants reach the six- to eight-leaf stage, temperatures can be dropped to as low as 39° to 43° F (4° to 6° C). About two weeks before flowering, when flower buds are visible but not before, raise temperatures to 52° to 57° F (11° to 14° C). Much research has been done concerning primula crop time and flowering. If plants are given a longer daylength and higher light levels, they can be grown at higher temperatures and shorter crop times. With higher light levels, primula can be finished at 58° to 60° F (14° to16° C) without lowering overall quality. In most cases the maximum should be around 58° F (14° C).

Fertilization

Plants should receive a balanced fertilizer throughout their growth. Primulas are low to moderate feeders, so use 100 to 150 ppm constant feeding. Avoid ammonium nitrate fertilizers during winter months, and increase potassium slightly to harden plants and improve flowering. Keep soil pH below 6.0 because plants are prone to become chlorotic from lack of iron and manganese.

Watering

As mentioned earlier, newly potted primula must never dry out, but once plants are well established they can be run a little drier to keep them more compact. Too much water will promote a longer leaf. It's best to water thoroughly on clear days so plants dry out before nightfall.

Pests and diseases

As far as pests and diseases, primulas don't require too much attention except at certain stages of growth. In the summer, watch out for cutworms. Later in the fall and winter, aphids may be present. Primulas don't have any significant disease problems except the possibility of botrytis if greenhouses are too wet, especially during flowering. Water early in the day and on days that you can dry out greenhouses. Some growers will ventilate late in the day to dry out greenhouses.

Primula for Late Spring Sales

Many trials and studies have been conducted to produce primula for late spring sales. This has been successful on both the acaulis and veris or polyanthus types of primula. It's possible to sow primula from September through the beginning of November. After germination keep temperatures at 68° F (20° C). Daylength must be a minimum of fourteen hours, and supplementary light is required when light levels are below 743 f.c. At approximately ten weeks from sowing, flowers are induced. You can achieve this by using plugs. Then plants can be held under light a while longer before planting them into their final containers.

When plants are potted, keep temperatures at approximately 60° F (16° C) so plants can become established and attain some size before lowering the temperature. If sown on November 14th, plants should finish in mid-April.

Finishing Primula Fast

Growers have also used supplemental lighting to finish primula faster and at a higher temperature of approximately 60° F (16° C). If they need a specific number of plants for a specific sales contract, they can speed up flowering by using supplemental lighting and not reduce the quality of the crop. Temperatures can actually be slightly higher in some cases depending on daylength and light intensity.

Fall Flowering Primula

Primula can be finished rather quickly in the fall. Some of the veris or polyanthus types can be finished in four and a half months from sowing. If sown on May 21 or in early June, they can be ready by mid-October. Finished veris or polyanthus primula will produce three to five flowers.

Tom Linwick is technical representative, Daehnfeldt, Duvall, Washington. June 1996.

Producing Perfect Primrose

Tom Linwick

As you head into the winter months, making cultural adjustments such as increasing light levels and changing fertilizer can greatly improve your primula quality.

Once plants are well established in their final containers, shade cloth isn't necessary unless high temperatures become a problem. You should have removed shade cloth by mid-September to early October to increase light levels and improve flowering.

Change your fertilizer to a balanced mix using calcium nitrate instead of ammonium nitrate and slightly more potassium nitrate. Watch the soil pH so it doesn't exceed 6.0 and cause iron and manganese deficiencies. Constant feeding at 100 to 150 ppm is recommended. If you need to green plants up, one feeding with an ammonium nitrate–type fertilizer will work within a week.

Botrytis can be a major problem in primrose production, particularly with plants in flower, but it can be controlled by proper watering. The best time to water is on sunny days, soaking soil completely early in the day so foliage dries out by nightfall.

Try to provide as much airflow as possible to keep humidity levels down and to control disease. This is difficult during the winter but greatly improves quality.

Plants' spacing in their final stages will make for less cleaning when they're ready for sale. If you're growing in 4-in. pots, spacing of four per square foot is adequate. Or some growers take the center row out of a flat of 4 in., allowing space between flats for better air circulation. You can grow primrose rather tight by watching watering and fertility closely to control plant size and development.

Tom Linwick is technical representative, Daehnfeldt Seed, Duvall, Washington. November 1995.

Growing *Primula acaulis*

Teresa Aimone

In June, begin sowing your *Primula acaulis* seed or plan your plug purchases. Before selecting varieties, know when you want your primula to bloom. There are early-, mid- and late-season varieties, and some have bloom times that aren't as easily manipulated by temperature and light as others.

Store seed at 45° F (7° C) (this is the temperature of your refrigerator). When you're ready to sow seed, allow it to warm up to room temperature before you open the package. Sow seed within one hour after opening the package so no unnecessary moisture condenses on it.

Optimum germination temperature is a constant 59° to 65° F (15° to 18° C). *Primula acaulis* doesn't like fluctuating temperatures. Don't let temperatures rise above 68° F (20° C). Germination takes ten to fourteen days. Maintain humidity as close to 100% as possible. Seeds don't require covering to germinate, but a light layer of coarse-grade vermiculite can help maintain humidity levels. You don't want seed to dry out at any time during germination.

Begin feeding at week four (end of Stage 2/beginning of Stage 3), using 25 to 50 ppm potassium nitrate. Maintain pH of 5.5 to 6.0 and EC of 1.25. Drop temperatures to 55° to 60° F (13° to 16° C) for week six through transplanting. (Transplanting begins when plants have at least three true leaves).

Teresa Aimone was a regional specialist Southeast, S&G Seeds, Coppell, Texas. June 1997.

Protea

❋

Producing Israel's Unique Protea

The Proteacea family is one of the oldest and lowest in terms of genetic development. About 75 genera and 1,350 species of small trees and shrubs, commonly found in Australia and South Africa, make up the family. Proteas have high value as flowering pot plants, bedding plants, cut flowers (fresh and dried), and fruit trees (macadamia is in the Proteacea family). They are mainly characterized by their evergreen nature, long flowering period of several months, and favorable flowering time (fall, winter, and the beginning of spring).

Though Europe has been the traditional protea market, in recent years, two markets—the U.S. and Japan—have expanded significantly. The worldwide market can still absorb large quantities of proteas without lowering prices, according to a recent study by the Israel Ministry of Agriculture. Still, the market is in need of new varieties to refresh existing selections.

Proteas are commercially grown primarily in the Southern Hemisphere and are found as commercial cut flowers in New Zealand, Australia, South Africa, Hawaii, and Zimbabwe. Outdoor production has also increased in the Northern Hemisphere. Here are some tips for successfully producing these unusual and increasingly popular plants.

Growing Conditions

Optimal climatic conditions for protea production are those prevailing in the temperate Mediterranean zones. Most proteas grow outside in full sun, but not in extremely high temperatures. Pruning can be done right after flowering, toward spring.

Soil is a more crucial factor; proteas usually prefer acidic soil with a 5.5 to 7.0 pH. However, some of the genera and species can be successfully grown with higher pH levels.

In Israel, *Leucadendron* is a very popular genus with a wide range of species. If you have alkaline soil, as in Israel, one way to keep pH down is to plant proteas on small mounds while keeping pH levels below 7 by adding sulfuric acid during the first year of production. Another strategy is to graft varieties on root stocks resistant to high pH, excess phosphorus, and soil problems such as phytophthora and nematodes. Grafting allows protea production in most types of soil in Israel, as long as they're well drained and have good airflow.

Proteas have two root systems: the usual roots for support and absorption and proteoid roots, which is a root system composed of lumps of small short-lived roots located close to the surface. As these roots are very efficient in absorbing minerals, any excess quantity is easily absorbed and may kill the plant.

Preplanting

As cut flowers grown in the open field, proteas require minimal preplanting preparation. There is no need for special structures such as greenhouses or shade houses. Buying plants is the main investment, but they are perennial stock, which can yield for at least ten years.

Harvest

Unlike plants such as carnation, chrysanthemum, and rose, proteas can be harvested over a long period of time at different stages of development. They can be harvested for a variety of uses (single, spray, or with flowers or greens). This means harvest time is flexible.

Pests

Proteas are attacked by few pests and diseases; therefore, spraying is usually minimal.

Postharvest

After harvest, machinery may be used for bunching and packing. Proteas have a long vase life—so long, in fact, that proteas can be exported from Israel by sea, rather than by air, significantly reducing overall costs.

Danziger "Dan" Flower Farm, Beit Dagan, Israel. December 1997.

Roses

❋

Producing Mini Pot Roses

Ron Ferguson

Though pot roses have been a commercially valued crop for many years, new varieties developed during the last 25 years have brought pot roses to the forefront as a highly sought-after product. Consumers are dazzled in many markets by the assorted colors, fragrant varieties and durability of pot roses.

Demand in the U.S. has centered primarily on 4-in. product; however, newly developed varieties have sparked interest in 5-in. and 6-in. potted varieties. The larger cultivars feature fully developed flowers with spiral opening buds.

Plugs and Liners

Liners are sold in forty-cell trays. You can expect at least three live cuttings per cell. The plants are about nine weeks old, rooted and have one cutback. The advantage to using liners is that no propagation expense or time is needed.

Prefinished

If ordering prefinished plants, you'll receive a 4- or 6-in. pot that has at least one cutback. It will be fully rooted and ready to develop new growth.

Finished

If you're buying finished plants ready to sell, be sure to acclimate them.

Medium

Medium is the base in any production program: The most commonly used mix for pot roses is a light peat-perlite with additives. It can be a peat-vermiculite mix or the new coconut "coir"-vermiculite. Many growers prefer not to mix. They use straight peat or straight coir. Adding superphosphate at ½ lb./cu. yd. is advisable. All versions should have a complete fertilizer such as 14-5-20 added at ½ lb./cu. yd.

Test the EC of the media by wetting, then squeezing enough to fill a 4-in. pot. Collect the water in a cup or beaker, and run the test. Desired EC is 1.5 to 1.8, pH 5.5 to 6.4. Limestone or dolomite may be added to raise pH. Coir normally has a pH of 6.0 to 6.2 and doesn't require adjustments with limestone.

Direct Sticking

The majority of pot roses are grown by direct sticking. Mowing previously planted pots is the method primarily employed to obtain cuttings.

Propagation must be done in a clean, disease-free environment. Be sure to tell your personnel to be careful in avoiding contamination of the cuttings. Use rubber gloves, disinfected tools, and bleach dips.

If you're beginning with stock plants, be sure to have them in 6-in. pots. Provide adequate spacing and optimum growing conditions. It's important to maintain only the best, most vigorous specimens.

Moisture loss is a constant concern. Avoid high temperatures in the handling areas. Pre-cool cuttings before trimming. Place wet cuttings in a plastic bag, and store overnight at 34° F (1° C). After cooling, trim cuttings to approximately 1 in. with a five-leaflet node. Rooting hormones can be used either as powder or liquid (IBA 0.3%); however, most varieties of pot roses will root well without hormones. Experiment to determine your best results. Coconut fiber is proving to be an excellent rooting medium.

Next, move the prepared cuttings directly to the propagation area. This area is filled with premoistened media in the final pots. You should use fog, high-pressure mist, plastic tents, and top and bottom heat. The best rooting temperature is 74° F (23° C). Bottom heat will bring the soil to that temperature.

Stick four cuttings per 4-in. pot. Six-in. pots usually have four cuttings around the edge and one in the middle. After watering in, drench with a fungicide. Use Terraguard or a mixture of Chipco 26019 and Subdue 2E at label rates.

Time to root will vary, but seven to ten days should be enough to have roots visible at the bottom of the pot. (Coir has been quicker to root by three days.) Adding CO_2 (1,000 ppm) and lights (800 f.c. or more) will improve the rooting process. Daylength can be extended to ten hours. Use supplemental lighting to maintain light levels at a minimum of 800 f.c. Using a light meter and electronic controls will accomplish this. Relative humidity should be maintained at 95 to 99% throughout the rooting process.

Check EC after rooting to see how much fertilizer has been leached. If the reading is below 1.5, apply fertilizer. Use 150 ppm nitrogen from a 20-10-20 fertilizer. Develop a consistent schedule for fertilization. A good recommendation is to fertilize twice and use plain water once.

After propagation, move the plants to an area with subirrigation using ebb-and-flow benches or watering mats. You can hand water, but this is difficult and doesn't always cover evenly. Overhead watering booms are used, but bottom watering is best.

Then, lower the day temperature to 75° F (24° C) and night temperature to 64° F (18° C). Watch root development throughout all stages of growth. Properly watered roots with good oxygen exchange will be thick and have root hairs. Roots that are thin and discolored with no hairs indicate too much water or fertilizer.

The first cutback should be after propagation, usually seventeen to twenty-one days. A cutter with a vacuum pickup is normally used. Most varieties will be ready, but occasionally some are slower. An after-pinch may be necessary a few days later as uncut shoots push up over the leaf canopy. Approximately twenty days later, perform the second cutback. Use cuttings from this and the first cut for new propagation.

Ten days after the second cutback, space the plants at least one pot width apart. After spacing, time to finish is twenty-five days.

Growth Retardants

Bonzi is the growth regulator of choice. The effect depends on concentration, amount of application, and frequency of application.

Begin sprays after the second cutback. Three sprays at 30 ppm, one week apart is a good guide. Because external factors can change the effects, test sprays are a wise practice.

Insects and Diseases

The pests of most concern are thrips and red spider mites. To control these, start with clean conditions, screening vents, sticky cards, and observation.

For thrips, use Orthene, Mavrik or neem oil. For mites, use Avid, Pentac, Sanmite, or neem oil.

One note: If fungus gnats have been a problem in the past, coir will make a difference. Coconut fiber doesn't support gnat development.

Powdery mildew and botrytis are controllable with sprays and environment. Spray Triforine alternating with Cleary's 3336 and Rubigan for powdery mildew. Use sulfur vaporizers as a residual control. Control botrytis with Chipco 26019 or Cleary's 3336. Monitor carefully during propagation.

Packing and Shipping

Your shipping method will determine packaging. Pot wraps and sleeves can be used, or customers may have special requests. Shipping and storage temperatures should be 40° F (4° C).

Never ship or store roses with apples, pears, or cantaloupes, as ethylene gas will adversely affect the entire plant. Also, watch areas where propane lifts are used.

Labeling

Labeling is important and is a legal requirement. Patented varieties require a special tag that has the patent number and originator. Each pot requires a label. Royalty fees are paid to the patent owner. Without these fees, new varieties wouldn't be developed.

Ron Ferguson is senior horticulturist, Jackson & Perkins, Medford, Oregon. March 1998.

Managing Rose Temperatures
for Better Buds and Fewer Blinds

Many Dutch growers are manipulating temperatures on roses to increase bud size, produce fewer blind shoots, fewer bull heads, and faster bud breaks, according to Roses Inc. They divide the temperature regime over a twenty-four-hour period into five periods. The key to the strategy: increasing the temperature by 37° to 39° F (3° to 4° C) and maintaining it until about two hours after sunset. This temperature period can vary somewhat depending on the total daily radiation.

After this elevated temperature period, they drop the temperature to 54° to 57° F (12° to 14° C) for two to three hours to obtain the required daily twenty-four-hour temperature average. It shouldn't last too long to prevent plants from reaching dewpoint. This is also when HPS lights are turned off during the winter. The ultimate goal with the temperature regime is to maintain the daily twenty-four-hour average at 63° to 66° F (17° to 19° C) during the winter and 70° to 75° F (21° to 24° C) during the summer, depending on the available amount of light. The higher the light sum, the higher the temperature, and vice versa. Remember, the daily light sum in the Netherlands is about 50% or less of Canadian/Michigan/New York conditions.

By doing this, sugars formed during the day are transported to flower buds or developing shoots rather than to roots. The critical time period for this transport process is at the end of the day and beginning of the night. Quicker bud breaking after a cut is one important effect. Roses may be a bit shorter, though you might expect the positive DIF to cause longer stems. Growers using this technique say the information was provided through private funded research by a group of twenty Dutch rose growers.

Culture Notes, May 1996.

Electrostatic Sprayers Provide
Best Mite Control on Cut Roses

In experiments on cut roses, University of California and California Flower Cooperative researchers have achieved better two-spotted spider mite control and longer residual activity with some chemicals by using electrostatic sprayers than by using conventional sprayers. "Avid's effectiveness was dramatically improved in the 1994 experiment with electrostatic sprayer applications, and we also saw improvement in the 1995 experiment," says Steve Tjosvold, University of California Cooperative Extension, Watsonville, California. Avid applied with an electrostatic sprayer had one of the best residuals, twenty-eight to thirty-three days. When applied with a conventional sprayer, Avid recorded only a twelve- to fifteen-day residual.

In 1995, researchers treated four replicates of thirty-six mature rose plants at a Watsonville, California, commercial greenhouse with conventional spray applications. They applied chemicals with a standard spray wand with a two-nozzled head at 400 gal./acre and 250 psi static pressure and made electrostatic spray applications with an Electrostatic Spraying Systems spray wand with a three-nozzled head at 20 gal./acre. Rates used were from chemical labels. They tagged leaflets to use as samples, and then selected leaflets in samples after plants were treated and counted spider mites on each leaflet.

Electrostatic applications improved the effectiveness of Avid and AC 303, 630, and experimental chemical (0.08 lb./acre). Other rates may produce different results. AC 303, 630 (0.08 lb./acre) and Avid treatments with electrostatic applications had the longest residual control—at least twenty-two days. The contact miticide Triact had the shortest residual action, lasting about four days. All other treatments lasted eight to fifteen days. In the 1995 experiment, researchers didn't observe phytotoxicity on any treated plants when they lowered the rates on some experimental products.

In similar experiments in 1994, researchers made conventional spray application treatments to four replicates of seventy-two mature rose plants with a standard spray wand with a two-nozzle head at 500 gal./acre and 250 psi static pressure. They made electrostatic spray application treatments with an Electrostatic Spraying Systems spray wand with a three-nozzle head at 23 gal./acre.

Experiment comparing electrostatic and conventional sprayers, 1995							
		Spider mites per leaflet (days after treatment[1])					
Spray method	active ingredient	0	4	8	15	22	28
	lbs./acre						
Conventional spray (400 gal./acre)							
AC 303, 630 2SC	0.04 lb.	13.76	18.06	19.18	22.18	39.31	48.30
AC 303,630 2SC	0.08 lb.	20.79	7.34	10.76	12.96	34.34	48.58
Pentac 4 F	1.00 lb.	18.32	18.92	23.72	12.89	22.85	33.64
Avid 0.15 EC	0.02 lb.	12.96	17.72	13.32	27.98	43.69	69.87
Sanmite 75 WP	0.50 lb.	13.18	11.02	11.90	6.76	20.25	38.19
Triact 90 EC	3.60 gal.	15.68	10.82	11.16	14.06	39.31	86.30
Electrostatic spray (20 gal./acre)							
AC 303, 630 2SC	0.04 lb.	28.94	10.30	15.37	19.27	29.16	51.55
AC 303, 630 2SC	0.08 lb.	17.14	2.16	10.24	6.71	5.15	20.16
Pentac 4 F	1.00 lb.	17.81	24.21	21.90	21.07	32.95	47.33
Avid 0.15 EC	0.02 lb.	20.88	8.01	8.58	9.55	8.12	22.85
Sanmite 75 WP	0.50 lb.	25.10	16.48	20.16	9.92	25.00	21.44
Triact 90 EC[2]	0.36 lb.	13.25	12.25	21.07	27.04	51.55	62.57

[1]Treatment date is October 19, 1995. Day 0 is pretreatment count on October 18, 1995.
[2]Active ingredient/acre rate for electrostatic application isn't equivalent to the conventional application rate, as is the case in other treatments.

Culture Notes, June 1996.

Rudbeckia

Rudbeckia Variety Review

Jim Nau

Rudbeckia hirta is an excellent perennial for well-drained areas in full-sun locations. Rudbeckias are often treated as annuals in the Midwest or in any area where the fall can be cold and wet. They don't appreciate overly moist conditions, and dwarf varieties often die from foliar diseases brought on by excessive moisture in combination with high humidity, such as powdery mildew. Plants flower readily from seed the first season, and varieties such as Marmalade, Goldilocks, Becky Mix, and Toto will even flower well in pots or packs.

Rudbeckias germinate in five to ten days at 70° F (21° C). The seed can be lightly covered during germination. For flowering 4-in. pots in May, sow in January, and use one plant per pot. For taller varieties, allow eleven to fifteen weeks for green packs.

Among varieties, both Marmalade and Goldilocks are dwarf selections that grow to only 15 in. tall in the garden. Marmalade does have problems with powdery mildew, but provides color from planting until August. Flowers are single, up to 3 in. across, in a golden yellow to light orange color. Goldilocks, which flowers ten days earlier, is a semidouble- to double-flowering variety that grows to 12 in. tall in the garden. Flowers range from 3 to 4 in. across, and the variety is excellent in packs or pots.

Becky Mix is a simple mix with two predominant flower colors—orange and orange red. Occasionally, a golden yellow color pops up. Becky is a pot plant variety growing only 10 in. tall in the pot and 12 to 14 in. tall in the garden. Flowers are single and are 2½ to 3½ in. across.

Indian Summer is an All-America Selections award winner with huge, 5- to 9-in., single to semidouble golden-orange flowers on plants 30 to 36 in. tall.

In the taller varieties, Double Gold is of particular value for cutting or as a background plant in the garden. Double Gold has double flowers up to 6 in. across in a stable color of golden orange.

Jim Nau is trials and new varieties manager, Ball Horticultural Co., West Chicago, Illinois. Excerpted from the 16th edition of the **Ball RedBook.** *September 1997.*

Schizanthus

Schizanthus: Success with Today's Short Varieties

Will Healy

Schizanthus isn't a new crop, but today's varieties are very different from those of the past, with short, compact habits, and shatter-resistant flowers. Older types were often tall and open, with flowers that shattered easily. Schizanthus has attractive, pastel-colored flowers that give it the common name "poor man's

orchid." The foliage is light green and fernlike. An excellent crop for winter and spring, schizanthus not only grows well at cool temperatures, but actually performs best at that time.

Temperature and Timing

Growth is affected by temperature: Flowering is fastest at higher temperatures. Plants have the most compact and uniform habit at lower temperatures. Grown at 47° F (8° C), plants require one 165 days to flower; those grown at 65° F (18° C) flower in 90 days. However, plants grown at lower temperatures were shorter with a better growth habit.

You can produce acceptable plants at either temperature, so you must weigh fuel costs against the costs of longer crop time. A 60° F (16° C) average temperature is a good compromise: Days to flower increased to one hundred, and plant habit was very acceptable.

In addition to temperature, light intensity also affects days to flower. At 53° F (12° C), schizanthus requires 143 days to flower at an average of 816 f.c., but only 125 days at an average of 1,084 f.c. Photoperiod doesn't appear to influence flowering.

Today's schizanthus doesn't require pinching to produce uniform plants. Pinched plants do have a somewhat more uniform habit, but days to flower is increased slightly.

Nutrition

Though no controlled experiments have been conducted, schizanthus appears to be a moderate feeder. Fertilization with a balanced fertilizer containing primarily nitrate nitrogen at 200 ppm has been successful.

Foliage is naturally light green, and deficiency of any mineral nutrient will accentuate this undesirable trait. No controlled studies have been conducted, but it's believed that magnesium sulfate (Epsom salts) at 1 lb. per 100 gal. applied as a drench just before flowering will improve foliage color.

Pests

No significant insect pests have been identified on schizanthus, and no phytotoxicity has been observed from any of the pesticides used. Monitor for thrips, whiteflies, and aphids, taking appropriate control measures if insects are observed.

Likewise, schizanthus is relatively disease-free. The most consistent problem is anthracnose-like leaf spot on lower foliage. Plants grown on ebb-and-flood benches have little or no leaf spot, while plants that are overhead-watered show symptoms. Sprays with Chipco 26019 appear to reduce the severity of the disease. Botrytis can also be a problem during periods of high humidity. A general recommendation for disease prevention is to avoid overhead watering, maintain ample air movement, and water early in the day to allow the foliage to dry.

Height Control

Today's new schizanthus varieties are naturally short and should be in proportion to their containers if grown well. They respond to applications of A-Rest, B-Nine, Bonzi, and Sumagic. Two sprays with 2,500 ppm B-Nine will result in shorter plants with little or no delay in flowering and darker green foliage, although B-Nine isn't registered for use on schizanthus.

Avoiding high temperatures, providing adequate light (less than 1,000 f.c.) and spacing plants so foliage doesn't overlap is usually enough to prevent excessively tall plants without using chemical regulators.

Postproduction Handling

Schizanthus has excellent pot-production longevity, with today's new varieties lasting two weeks with minimal care in a home or office. Schizanthus can withstand three days of shipping in a closed box and remain salable, although shipping does reduce the expected longevity of the plant by a few days.

Will Healy is manager, Technical Services, Ball Horticultural Co., West Chicago, Illinois. November 1997.

Schlumbergera (Holiday Cacti)

Schlumbergera Success Tips

Thomas H. Boyle

Holiday cacti, *Schlumbergera* species, have become a staple pot crop for specialty sales. Christmas cactus, *Schlumbergera* x *buckleyi*, has purplish-brown anthers, ribbed ovaries, and leaf segments with rounded margins. In contrast, Thanksgiving cactus (*S. truncata*) has yellow anthers, rounded ovaries with no ribs, and segments with prominent teeth on the margins.

Growth from Cuttings to Transplanting

Plants are propagated by rooting mature, single-segment cuttings obtained from vegetative stock plants. Remove cuttings from stock plants by twisting the segment

180 degrees and pulling it upward. Take cuttings from the top of the stock plants, and use only the mature terminal and subterminal segments for propagation. Avoid taking cuttings near the base of stock plants because they're more likely to be contaminated with soil-borne pathogens.

Propagate cuttings in cell packs or 1½- to 2-in. pots using two to four cuttings per cell or pot. Propagation media should be free of pathogens and well-drained. Keep propagation media temperature at 70° to 75° F (21° to 24° C) during rooting. Segments will root equally well using intermittent mist, high-humidity tents, or periodic hand-watering, as long as media remain moist and warm. Strict sanitation will reduce diseases and minimize the need for fungicides.

Propagate cuttings between November and March for sales the following November and December. Cuttings propagated during naturally short days (SD) from early September until late April should be given long-day (LD) photoperiods to promote vegetative growth. LD photoperiods can be provided using "night-break" lighting from 10 P.M. to 2 A.M. at 5 to 10 f.c. minimum light intensity at plant level.

A newly propagated cutting will often produce only one new shoot, resulting in a sparsely branched plant. If new shoots are pinched off at about six to eight weeks after sticking cuttings, then multiple shoots (two to four) will develop. Pinching will result in a fuller, higher quality plant.

Cuttings propagated between November and March will be ready for transplanting in April, May or June. Use one cell per 3½- to 4-in. pot, three cells per 5- to 6-in. pot, four cells per 6-in. hanging basket, and eight to ten cells per 8-in. hanging basket. Plants can be grown at pot-tight spacing during most or all of the growing period.

Culture

Use a growing medium that's well-drained, pathogen-free, and adjusted to a pH of 5.7 to 6.5. Many growers use a commercially formulated soilless mix that's composed primarily of sphagnum peat.

In general, holiday cacti will tolerate underwatering better than overwatering. Saturating growing media for prolonged periods will predispose root systems to attack by soilborne disease organisms.

High-quality plants can be produced using a balanced N-P-K fertilizer with micronutrients and applying 150 to 200 ppm nitrogen at each watering. Some growers use calcium nitrate and potassium nitrate to supply about 180 ppm nitrogen, 390 ppm potassium, and 53 ppm calcium at each watering, also applying a balanced N-P-K fertilizer with micronutrients once a month at 150 ppm nitrogen. High-quality plants can be produced using either nitrate or ammonium as the nitrogen source. Plants grown in soilless media will benefit from periodic applications of magnesium sulfate (Epsom salts) at 20 oz. per 100 gal., giving about 150 ppm magnesium. Begin fertilization as soon as roots develop on newly propagated cuttings.

The pH of growing media should be kept above 5.5. Plants will take up high amounts of iron and manganese when the pH drops below 5.5, leading to serious plant damage. Avoid using fertilizers with strongly acid reactions.

Under optimum growing conditions, holiday cacti will produce one tier of growth every six to eight weeks. Greenhouse temperatures should be maintained at 62° to 65° F (17° to 18° C) nights and 68° to 72° F (20° to 22° C) days. Provide ventilation above 74° F (23° C). Maintain light intensity at 1,500 to 3,000 f.c. on a year-round basis.

Flowering

Holiday cactus is an SD plant when grown at temperatures ranging from 60° to 75° F (16° to 24° C). The critical day length (the photoperiod separating SD from LD responses) is between twelve and a half and fourteen hours for plants grown at 64° to 65° F (18° C) nights and 70° to 72° F (21° to 22° C) days. Growers can use natural flowering or controlled flowering for producing holiday cactus.

Natural flowering

In holiday cactus, the long nights trigger flowering, not the short days. In the northern U.S., natural flowering will occur primarily in mid-November when plants are grown under natural day lengths and 62° to 65° F (17° to 18° C) nights (Table 1). Plants will flower earlier, though, if night temperatures dip into the 50s during August and early September. In the southern U.S., natural flowering will occur mainly in early to mid-December.

Table 1. Natural flowering times of holiday cactus varieties

Cultivar	Mean date of flowering*				Range (days)
	1988	1989	1990	1991	
Christmas Charm	Nov. 12	Nov. 17	Nov. 7	Nov. 15	10
Christmas Fantasy	Nov. 18	Nov. 24	Nov. 12	Nov. 18	12
Gold Charm	Nov. 14	Nov. 25	Nov. 18	Nov. 17	11
Lavender Doll	Nov. 21	Nov. 25	Nov. 18	Nov. 21	7
Lavender Doll II	Nov. 27	Dec. 5	Nov. 30	Nov. 25	10
Rocket	Nov. 14	Nov. 18	Nov. 5	Nov. 11	9
White Christmas	Nov. 15	Nov. 24	Nov. 14	Nov. 19	10
Mean of all seven cultivars	Nov. 17	Nov. 20	Nov. 15	Nov. 18	

Note: Trialed over a four-year period at the University of Massachusetts, Amherst. Greenhouse temperatures were maintained at 65° F (18° C) nights and 70° to 72°F (21° to 22° C) days.
*First flower open. Each mean represents five to eight pots.

As a method, natural flowering has two main disadvantages. Each variety has a limited flowering period; thus, several varieties are needed to provide a continuum of flowering plants. Also, crop scheduling is difficult because the exact time of flowering varies from year to year (Table 1).

Controlled flowering

Holiday cactus can be scheduled on a year-round basis by using controlled flowering, which requires accurate control of temperature and photoperiod. Maintain greenhouse temperatures at 62° to 65° F (17° to 18° C) during the night and 68° to 72° F (20° to 22° C) during the day. During natural LD (late April to early September in the northern U.S.), flowering can be induced by reducing daylength to eight or nine hours daily (giving sixteen or fifteen hours of continuous darkness, respectively). Maintain SD conditions on a daily basis for at least three weeks. Take proper precautions to prevent high temperatures under black cloth; poor or uneven budset may occur if the temperature exceeds 75° F (24° C) during SD.

Flower bud development is accelerated as the average daily temperature (ADT) increases from 59° to 75° F (15° to 24° C) (Table 2). The information in Table 2 can be used to schedule the time of flowering. Measure the lengths of the largest buds on several plants, and calculate the average bud length. Find the number in Column 1 that corresponds to average bud length, then read across to determine how many days are required for the buds to open. For example, if the largest buds averaged 8 mm in length, approximately 35, 25, 19 or 16 days would be required for these buds to open if the ADT were kept at 59°, 65°, 70°, or 75° F (15°, 18°, 21°, or 24° C), respectively.

Table 2. Temperature, bud length, and flowering of holiday cactus

Bud length	Days to flowering at various daily temperatures			
(mm)	59° F/15° C	65° F/18° C	70° F/21° C	75° F/24° C
1	70	50	39	32
2	58	41	32	26
3	51	37	28	23
4	47	33	26	21
5	43	31	24	19
6	40	28	22	18
7	37	27	21	17
8	35	25	19	16
9	33	24	18	15
10	31	22	17	14
11	30	21	17	14
12	28	20	16	13
13	27	19	15	12
14	26	18	14	12
15	25	18	14	11
16	24	17	13	11
17	23	16	13	10
18	22	15	12	10
19	21	15	12	9
20	20	14	11	9
25	16	12	9	7
30	13	9	7	6
40	8	6	5	4
50	5	3	3	2
60	2	1	1	1

Note: Camilla, Dark Marie, and Madisto varieties under four different average daily temperatures.

Delaying flowering with long days

Growers in the northern U.S. often find that some holiday cactus varieties flower too early for Christmas sales when plants are grown under natural photoperiods. Flowering can be delayed by maintaining plants under LD starting in the first week of September. Use "night-break" lighting from 10 P.M. to 2 A.M. (about 5 to 10 f.c. at top of plants) to prevent flowering under natural SD. Incandescent lamps are most commonly used for providing LD, but high-pressure sodium lamps are also effective. Temperatures must be maintained above 60° F (16° C) and preferably from 62° to 65° F (17° to 18° C) during LD photoperiods. Lighting from sunset until about 10 P.M. will also keep plants vegetative, but night-break lighting is more cost-effective.

Thomas H. Boyle is associate professor, University of Massachusetts, Amherst. August 1997.

Holiday Cacti
Teresa Aimone

Two types of holiday cacti (zygocacti) are grown for late fall/early winter sales: *Schlumbergera bridgesii* (Christmas cactus) and *Schlumbergera truncata* (Thanksgiving cactus). Both bloom around their respective holidays. Christmas cacti have narrow stem segments or pads and smoother margins. Thanksgiving cacti are noted for their deeper, serrated margins and wide pads.

Prune plants in June for fuller finished plants. At an average daytime temperature of 65° to 70° F (18° to 21° C), it takes about four to six weeks to put on one new pad of growth, so keep that in mind when pinching plants. Finished holiday cacti look best when plants are one-and-a-half to two times taller than the container height. Higher temperatures will prompt faster growth, but they can also cause chlorotic foliage, thinner pads or bud drop. If light levels are kept below 2,500 f.c., plants can tolerate temperatures above 90° F (32° C).

Fertilize holiday cacti with 150 to 200 ppm nitrogen; you can use nitrate or ammoniacal forms of nitrogen. Maintain pH of 5.5 to 6.0, and keep salt levels below 3.0.

Teresa Aimone was a regional specialist Southeast, S&G Seeds, Coppell, Texas. June 1997.

Shamrock

❧

No Luck Required for Successful Shamrocks

Laurie Beytes

As a former grower, I can honestly say that there is no quicker and easier crop to grow than shamrocks for St. Patrick's Day. With low input costs and practically no labor or pest problems, they're almost a guaranteed sale for a much overlooked holiday. Here are the details on culture for *Oxalis regnellii*.

Scheduling

St. Patrick's Day is always on March 17. Blooming shamrocks will be ready for sale eight to ten weeks after potting when grown at 55° F (13° C) and six to eight weeks at 65° F (18° C). However, shamrocks can be marketed once foliage is fully expanded, which can be in as little as four to five weeks. Schedule your planting date accordingly.

Potting

Shamrocks are grown from small, scaly rhizomes. Plant six to eight rhizomes per 6-in. pot and three to five rhizomes per 4-in. pot, depending upon how full you want the finished plants to be. Eight rhizomes per 6-in. pot gave us beautiful, full, florist-quality plants. Lay rhizomes flat, ½ to 1 in. below the surface of any good, well-drained media. While shamrocks can be started pot to pot, spread them out for finishing for the best results. Six-in. pots will finish nicely on 9-in. centers. If you can't pot them up right away, store rhizomes at 41° to 45° F (5° to 7° C).

Irrigation and Fertilization

Keep media moist but not soggy. Once plants emerge, fertilize weekly with 200 ppm of liquid 20-20-20 or use one application of Osmocote 14-14-14. Maintain a pH of 6 to 7.

Temperature and Light

Shamrocks can be grown between 55° and 75° F (13° and 24° C); however, 60° to 65° F (16° to 18° C) is ideal. Keep light levels at 2,500 to 3,000 f.c.

Growth Regulators

Some growers may find it necessary to use a Cycocel spray at 75% of the normal rate to shorten leaf petioles and help green up plants. It may, however, delay flowering. We found that with a good fertilizer program and the high light conditions we had in Florida, growth regulators weren't necessary.

Pests

Although insects are rarely a problem on shamrocks, they can be susceptible to spider mites. Control with Pentac 50WP at 8 oz. per 100 gal. of water, two applications per week for two weeks. Diseases aren't a concern.

Varieties

Everblooming shamrock *O. regnellii* is the most common variety grown for St. Patrick's Day. It has green, three-lobed leaves and small white flowers. Other varieties with different foliage and flower colors are available, including:

O. bowiei has rose-pink flowers and green foliage.

O. hirta has pink flowers and moss-like foliage.

O. deppei Iron Cross has deep rose flowers and a red cross in the center of foliage.

O. latifolia has burgundy foliage and rose pink flowers.

O. purpurea Grand Duchess has white, pink or lavender flowers and green foliage.

O. maritiana has pink flowers and variegated green foliage.

Laurie Beytes is a horticulturist from Naperville, Illinois. December 1995.

Snapdragon

Snapdragons: Formula for Success

Linda Laughner and Brian Corr

Snapdragons continue to generate increasing interest among growers and consumers alike. Ensure that the popular crop is a profitable part of your cut flower program with this complete guide to quality snaps.

Seedling Scheduling

Stage 1—Sowing to radicle emergence (six to eight days)

Maintain 65° to 75° F (18° to 24° C) soil temperatures. Keep soil evenly moist but not saturated. Don't cover seed. Light isn't necessary for germination until the radicle emerges. Keep soil at pH 5.5 to 5.8 and soluble salts (EC) less than 0.75 mmhos/cm (2:1 extraction). Snapdragons are very sensitive to a high starter charge in mix. Keep ammonium levels less than 5 ppm.

Stage 2—Stem and cotyledon emergence

Maintain 65° to 75° F (18° to 24° C) soil temperatures. Reduce moisture levels once the radicle emerges. Keep soil evenly moist but not saturated for best rooting. Light levels should be between 450 and 1,500 f.c. Keep soil pH at 5.5 to 5.8 and EC less than 0.75 mmhos/cm. Maintain water alkalinity at 60 to 100 ppm. Begin fertilizing with 50 to 75 ppm nitrogen from calcium nitrate– and potassium nitrate–based fertilizer once cotyledons are fully expanded. Liquid fertilization may not be necessary at this stage if you incorporate sufficient nutrition in the growing media before planting. Irrigate early in the day so foliage is dry by nightfall to prevent diseases.

Stage 3—Growth and development of true leaves

Maintain 62° to 65° F (17° to 18° C) soil temperatures. Allow soil to dry thoroughly between waterings, but avoid wilting. This will produce the best root growth. Increase light levels to 1,000 to 2,500 f.c. Maintain soil pH at 5.5 to 5.8 and EC less than 1.0 mmhos/cm. Increase feed to 100 to 150 ppm nitrogen from 20-10-20, alternating with 14-0-14 or another calcium nitrate and potassium nitrate fertilizer. Supplement with magnesium one to two times during this stage, using magnesium sulfate (16 oz./100 gal.) or magnesium nitrate. Don't mix magnesium sulfate with calcium nitrate, as a precipitate will form. Occasional leaching with clear water helps to reduce soluble salts. Try to maintain a ratio of approximately 3 potassium: 2 calcium: 1 magnesium in media. Avoid ammonium-based fertilizer if growing below 65° F (18° C). Apply fungicides at the lowest recommended rate to control pythium, rhizoctonia, and thielaviopsis.

Stage 4—Plants ready for transplanting or shipping

Maintain 60° to 62° F (16° to 17° C) soil temperatures. Allow soil to dry thoroughly between waterings. Maintain soil of pH 5.5 to 5.8; EC less that 1.0 mmhos/cm for transplanting, less than 0.75 mmhos/cm for shipping. Fertilize with calcium nitrate– and potassium nitrate–based fertilizer as needed. Don't use ammonium-based fertilizer.

Transplanting and Fertilization

Snapdragons grow best in growing media that allow adequate aeration to the roots yet hold a steady supply of moisture. Test media before planting. Fertility should be moderate, between 1.0 and 1.75 mmhos EC with less than 10 ppm ammonium nitrogen. Media pH should be between 5.5 and 6.5.

Snapdragons are often described as "light feeders," yet no crop can grow well with inadequate nutrition. Growers usually incorporate phosphorus and calcium into growing media before planting and supply other nutrients with a soluble fertilizer during growth. Superphosphate incorporated at 5 lb. per 100 sq. ft. should supply sufficient phosphorous for the entire crop, except in very porous media. If media tests show calcium is low, incorporate limestone (if the pH is too low), or gypsum (if pH is acceptable), both at 5 lb. per 100 sq. ft. If you use phosphoric acid to modify water alkalinity, you may not need superphosphate applications.

Plugs are generally ready to transplant four to five weeks after sowing. When the second true leaves unfold, transplant plugs on a spacing of 10 to 12 plants per sq. ft., decreasing to eight plants per sq. ft. in seasons with low light. When buying in seedlings or plugs, allow seedlings twenty-four hours to acclimate to greenhouse conditions, then transplant promptly. Delayed flowering and loss of final product quality occur when seedlings are kept too long in plug trays. If holding is unavoidable, store plugs at 36° to 39° F (2° to 4° C) under fluorescent lights at 250 f.c. fourteen hours per day. Treat with fungicide before storage to prevent botrytis.

Irrigate seedlings with clear water after transplanting. Begin fertilizing at the next watering, using a well-balanced, low-ammonium (less than 40%) fertilizer at a rate of 150 to 200 ppm nitrogen. Constant fertilization with occasional clear water leaching can be used until the flower buds show color and swell. Once flower buds show color and swell, use clear water only.

Insects and Disease

Snapdragons are relatively pest-free in comparison to many crops. Aphids, mites, and thrips are the most common pests on mature plants, while seedlings are damaged by fungus gnats and shorefly larvae. Control these pests by excluding them from the growing area. Scout the growing area frequently to find infestations before they become severe. Nicotine sulfate, vapona, and malathion have been reported to cause phytotoxicity of snapdragons and should be avoided. Chlorpyrifos and some other insecticides can damage budded spikes. If the crop is free of insects going into flowering, then insecticides can usually be avoided when buds show color.

Fungal diseases that affect snapdragons include downy mildew, powdery mildew, pythium, botrytis, rust, phyllosticta blight, and anthracnose. Downy mildew affects snapdragons as seedlings with stunting chlorosis and downward curling of leaves and on more mature plants as fuzzy, white growth on leaf undersides. Sterilized media, careful watering, heat, and ventilation help prevent it. Powdery mildew affects seedlings as well as mature plants causing white powdery growth on both leaf surfaces that left untreated can destroy lower leaves and spread to the upper plant parts, eventually causing white, circular blemishes on flower petals. Pythium doesn't kill snapdragons at low levels but decreases their vigor, resulting in uneven stem lengths, weaker flower stems, and poor quality, shorter flower spikes. Botrytis causes tan-colored stem lesions that can encircle stems, causing wilting of upper plant parts. Sanitation is essential in controlling botrytis, as the fungus persists in plant debris. Snapdragon rust causes faint, yellow spots on upper leaf surfaces and rusty-brown, circular pustules on lower surfaces. Keep water off foliage as much as possible to limit infection. Phyllosticta, primarily a problem on snapdragons in hot, humid areas, begins as brown or black foliar spots that enlarge to form light brown lesions dotted with small black fungal fruiting bodies. Lesions can also occur on stems, cracking or girdling the stem and causing the plant or branch to wilt. Anthracnose causes leaf or stem spots that are grayish white, sunken and form oblong areas with dark, narrow borders. Infected leaves usually die and drop. Tomato spotted wilt virus (TSWV-L) and impatiens necrotic spot virus (INSV) symptoms of brown or black stem lesions often don't appear until just before bloom. The best method of prevention is to monitor and control western flower thrips, which transmits these viruses, especially during seedling and young plant stages.

Scheduling

Snapdragon growth and flowering response depend on the interaction of light quality, light duration, temperature, CO_2 levels, humidity, soil type, and other environmental factors. Once flower initiation has occurred, night temperature has the greatest influence on flowering time and final quality. Varieties are separated into four groups based on their optimal growing conditions:

• Group 1—short days, low light, night temperatures 45° to 50° F (7° to 10° C).

• Group 2—short days (but not as short as Group 1), moderate light, night temperatures 50° to 55° F (10° to 13° C).

• Group 3—medium to long days, moderate to high light, night temperatures 55° to 60° F (13° to 16° C).

• Group 4—high light, long days, night temperatures higher than 60° F (16° C).

Generally, lower temperatures in these ranges give the best quality but at the expense of a longer crop time. The lower temperature is advisable during extended periods of low light.

Growers often say the fall transition from Group 3 to Group 2 is the most difficult time to schedule a continuous succession of quality snapdragons. Excessively warm temperatures and high light at the young plant stage (late summer) can make Group 2 snapdragons bloom too early and too short. On the other hand, unusually cool nights, even after flowers have initiated, can drastically lengthen the crop time of Group 3 varieties. Intermediate varieties (Group 2, 3) are excellent choices for harvest during this period. Alternatively, use the variety descriptions to choose varieties that help connect Group 2 to Group 3. This logical progression as daylight decreases is: Group 3, early Group 3, late Group 2, Group 2.

Snapdragon scheduling guide

| Group | North[1] | | | | | South[1] |
	Sow (week)	Transplant (week)	Flower[2] (week)	Sow (week)	Transplant (week)	Flower[2] (week)
1	33 to 35	37 to 39	50 to 7		Not applicable	
2	37 to 49	40 to 1	8 to 19	34 to 51	38 to 4	49 to 17
	30 to 32	34 to 38	44 to 49			
3	50 to 11	2 to 15	20 to 26	28 to 33	32 to 37	40 to 48
	25 to 28	28 to 33	37 to 43	2 to 10	5 to 14	18 to 24
4	13 to 23	16 to 27	27 to 36	11 to 26	15 to 31	25 to 39

[1]The North and South areas are separated at the 38th parallel, extending from San Francisco in the West through Kansas City to Washington, D.C. in the East.
[2]Regional conditions vary. Sowing, transplant, and harvest dates given above are general guidelines only.

You can grow high quality snapdragons using supplemental HID lights. In this production method, Group 3 snapdragons can be grown all year by lighting the plants when natural day lengths are less than twelve hours. Groups 1 and 2 aren't recommended for HID culture because they initiate flowers too quickly, which causes a short, weak stem. A light duration of twelve to fourteen hours per day and 350 to 400 f.c. of supplemental light is essential. Optimize conditions by increasing fertilizer to 300 to 500 ppm nitrogen, increasing temperatures to 60° to 62° F (16° to 17° C) at night and by adding CO_2 at 800 to 1,200 ppm.

Linda Laughner is a snapdragon breeder and Brian Corr is in international sales, PanAmerican Seed, West Chicago, Illinois. September 1996.

Trailing Snapdragons for Hanging Baskets

Terri Woods Starman and Millie S. Williams

In spring 1997 in trials at the University of Tennessee, we grew Lampion trailing snapdragons in 8- and 10-in. hanging baskets to gauge their performance and compare colors to find out how many plugs are needed for successful hanging

baskets. We trialed six varieties: Salmon-Orange, Purple, Yellow, Pink, Appleblossom, and White.

We received plugs from Ball FloraPlant May 1, 1997, and transplanted them into hanging baskets containing soilless peat-lite medium. We watered them in thoroughly and drenched with a fungicide to prevent root rot. A 30% shade cloth was pulled over the plants for several days after transplanting to prevent wilting while they established. A constant liquid fertilization program was used, alternating 20-10-20 with 15-0-15, depending on weekly EC and pH readings. We used 100 ppm nitrogen during the first two weeks, then 200 ppm nitrogen for the remainder of the experiment.

We grew plants in a 70° F (21° C)/65° F (18° C) (venting/night temperature set points) glasshouse. Baskets were grown on benches and spaced when needed so plants weren't touching. Plants were pinched once by cutting them to the edge of the pot when they were rooted out and shoots reached the edge of the container, approximately three weeks after transplanting.

We had an unusually cool, late spring in Knoxville. Botrytis occurred in some of our plants but was kept under control by careful attention not to overwater and with good air circulation. We experienced no major insect pests. Plants grew very well, and most cultivars reached full flower by eight weeks. After peak flower, flowers faded, and plants needed to be cut back. Purple and White continued to bloom longer than the others and were still in bloom in August.

Ten-inch Baskets

When grown in 10-in. baskets, all varieties took seven to eight weeks to finish. Plants were considered finished when the media was covered with foliage, the foliage was beginning to cascade over the side of the container, and flowers were evenly distributed over the top of the plant canopy.

We experimented with how many plugs were necessary to make a full, high quality basket by planting two, three, four, or five plants per 10-in. basket. The number of plugs needed differed by variety (Table 1). Yellow and Salmon-Orange were the most vigorous growers. Yellow was the most sturdy and quickly filled in the center of the basket. Three plugs were enough to fill the basket; four plugs looked even better. Five plugs weren't necessary for these two cultivars.

Appleblossom, Pink, and White were intermediate in growth habit. Three plugs weren't enough to fill a 10-in. basket; four plugs were sufficient to produce a full basket, but five plugs are recommended to make a higher quality basket. With less than five plugs per basket, flowers began to get old before there was full foliage coverage in the center of the basket.

Lastly, Purple was more spindly in habit, and even five plugs per basket didn't fill in well. An additional pinch could possibly improve this variety's appearance.

Eight-inch Baskets

We used two or three plugs per 8-in. basket, and it took seven to eight weeks to finish a full basket to have foliage covering the media and flowers distributed over the top of the foliage. Again Yellow was a winner, looking good in the smaller container. Using Yellow or Salmon-Orange, you could get by with only two plugs per basket, but three made a better quality basket.

Appleblossom looked good with three plugs per basket, but Pink and White looked rather sparse with only three plants. Purple formed flowers around the edge of the pot but never filled in the middle; therefore we concluded that this variety may need a second pinch or would possibly make a better companion plant than used alone.

We think the Lampion trailing snapdragons will be a fine addition to spring flowering basket lines. They'll also be useful as companion plants in spring color bowls because of their fine texture, pendulous habit and burst of color.

Table 1. Number of plugs per basket needed for each cultivar of Lampion trailing snapdragon.

Cultivars	Number of plugs per basket	
	10-in.	8-in.
Yellow	3 to 4	2 to 3
Salmon-Orange	3 to 4	2 to 3
Appleblossom	4 to 5	3
Pink	4 to 5	4
White	4 to 5	4
Purple	less than 5*	less than 4*

* Two pinches could reduce the need for additional plants.

Terri Woods Starman, associate professor of floriculture, and Millie S. Williams, graduate research assistant, University of Tennessee, Knoxville. February 1998.

Spathiphyllum

Spathiphyllum: Success for Every Market

Gary R. Hennen and Steven E. Hotchkiss

During the last twenty years, spathiphyllum's long-lasting, showy, white flowers and ease of growing have helped it gain popularity in the tropical foliage industry. Available in 3- to 14-in. pot sizes, it's a very durable plant for consumers.

Young spathiphyllums are available year-round from tissue culture or seed. Tissue culture offers growers the advantage of selected, named varieties, and improved crop uniformity. Spathiphyllum produced from seed has, until recently, lost volume to tissue culture production. Seed production, although economical, tends to lack the quality and uniformity demanded by growers, especially in larger pot sizes.

Potting Media
Spathiphyllum requires a potting mix with good drainage and water-holding capacity. A 1:1:1 ratio of peat, perlite, and bark is a good potting mix for the southern U.S., while coarse peat moss is common in Europe. Maintain pH at 5.8 to 6.5.

Nutrition
N-P-K ratio of 3:1:2 applied as a slow-release or liquid feed produces high-quality plants. Slow-release dry fertilizers, constant-feed liquid fertilization, or combinations of both are equally effective methods of applying nutrients. Many growers

incorporate slow-release fertilizer in the potting mix, supplementing later with liquid or dry applications. Some also use a weekly foliar nutrient spray of 1 lb. urea plus 1 lb. potassium nitrate plus 1 lb. magnesium nitrate per 100 gal.

Watch for nutritional deficiency symptoms. Magnesium deficiency appears as golden-yellow margins on lower leaves. Preventing deficiencies with supplemental magnesium is much more effective than trying to reverse a deficiency. Iron and manganese deficiencies show up as reduced growth rates and chlorotic leaves and are common during winter months when soil temperature is below 65° F (18° C). Sulfur deficiencies, seen as overall chlorosis of foliage, are sometimes found when using highly refined, low-sulfur fertilizers. Boron deficiencies can cause longitudinal ribbing of leaves, often on new growth. Potassium deficiencies can cause small yellow spots or "flecking" on lower leaves.

Watering

Irrigation frequency should be designed to keep soil media evenly moist during all phases of the crop cycle. Spathiphyllums easily tolerate overhead irrigation and do exceptionally well with drip or ebb-and-flow systems. They don't tolerate saturated soil conditions for extended periods of time. Overwatered spathiphyllums can have wilted or collapsed leaves, necrosis along leaf margins and extensive root damage.

Light

Production light intensities are somewhat cultivar dependent, although a range of 1,000 to 2,500 f.c. is commonly used. Plants grown in the lower f.c. range tend to have longer petioles, reduced branching, softer appearance, and darker green color. Under higher light intensities, plants tend to be more compact, branch more, and be lighter in color.

Temperature

The optimum temperature range for spathiphyllum is 65° F (18° C) nights and up to 90° F (32° C) days, but plants tolerate lows of 45° F (7° C) and highs of 95° F (35° C). Spathiphyllum won't tolerate frost or freezing temperatures without foliar damage. Plants grown at temperatures above 95° F (35° C) can suffer narrow leaves (strap leaf), color loss, inhibited root development, and reduced flower quantity and quality.

Diseases

A root rot caused by the soil-borne fungus *Cylindrocladium spathiphylli* can affect every spathiphyllum. Spread in soil and water, it can infect and kill very rapidly. The first symptom is lower leaf yellowing, which is sometimes accompanied by slight wilting that progresses to severe wilting. Splashing water can carry spores onto foliage, resulting in elliptical brown spots on leaves and petioles. Lower portions of petioles frequently rot, and at the final stage, roots are severely rotted and foliage totally collapses.

To combat cylindrocladium, you should always use pathogen-free plants from tissue culture or seed sources, sterilized potting medium and new pots. Frequently rogue crops and promptly remove infected or suspect plants from production areas. Always discourage bringing finished plants from other growers into your facility. Growing plants on raised benches or blocks is also effective in areas where the disease is established. Use Terraguard 50WP for effective control.

Growth Regulators

Two growth regulators are commonly used on spathiphyllum: benzyladenine and gibberellic acid. Benzyladenine is generally used at the young plant stage, but some growers also apply benzyladenine shortly after young plants have been planted into larger pots to enhance branching and fullness. Benzyladenine can be sprayed or drenched at 250 to 1,000 ppm. Treatments can inhibit root development if applied before roots are well established. Gibberellic acid is used extensively to force early or year-round flowering. With maturity, spathiphyllum will naturally flower consistently in spring and sporadically during the rest of the year. Since the market demands that spathiphyllum be sold with flowers, growers use gibberellic acid to gain a year-round sales advantage. It also helps program crops for holidays and promotions or weekly orders. Standard treatment is a single foliar spray of 150 to 250 ppm gibberellic acid, eight to fifteen weeks before sale. Treated plants may have narrowed new leaves, petiole stretching, and distorted flowers.

Applying growth regulators may not be legal in all growing areas. Many growers are concerned that currently labeled growth regulators will eventually be eliminated or their application severely restricted. Breeders are aware of these concerns and have started to develop varieties with improved branching and flowering.

Production Methods

Most growers use ten- to fourteen-week-old plants to line out larger pot sizes. Young plants from tissue culture (microcuttings) or seed are usually grown and delivered to finish growers in cell pack trays as small as 200 cells per tray and as large as 38 cells. Generally, tissue-cultured clumps produce very full plants but can lack uniformity as a finished product. They are very useful for small pot (less than 6 in.) production where growing times and chemical flower induction don't allow time for natural branching. Young plants produced from individual microcuttings tend to be more uniform, and most cultivars will produce full plants if given adequate time. Grower's needs or market requirements determine the finish grower's choice of the young plant cell size and plant material.

Growing Times

Growing times are directly related to cultivar, pot size, starter plant, and cultural environment. Generally, a 3- to 4-in. pot requires three to five months, a 6-in. pot seven to nine months, an 8-in. pot nine to eleven months, a 10-in. pot ten to twelve months, and a 14-in. pot sixteen to twenty months. Growers should consult with their young plant supplier for specific cultivar growing times.

Production and Marketing

Spathiphyllum growers, particularly those supplying large volume mass market accounts, must work closely with young plant producers to develop a realistic production program. Due to tissue culture economics and production constraints, regular production schedules and effective planning are a necessity. Most growers can't consistently grow and promote spathiphyllum for peak demand periods or promotions without having it as part of their year-round product mix.

Periodically during the last few years, the supply of several nonproprietary or "generic" spathiphyllum (Petite and Viscount) has exceeded market demand, resulting in reduced prices for starter and finished plants. The oversupply was primarily caused by extensive Third World tissue culture production and growers' willingness to absorb overproduction at attractive prices. Unfortunately, once the oversupply hit the finished market, sale price reductions reduced profit margins below any savings achieved from cheaper starter plants. To reduce the impact of oversupply, growers should consider proprietary plants protected by plant patents and trademarks as part of their product lines. Proprietary plants are generally controlled by a limited number of suppliers and offer more market stability.

To regularly market any quantity of spathiphyllum successfully, you must have a real dedication to competition. Clearly define your markets by considering how your buyers define quality standards for pot sizes, plant height, fullness, flower count, and overall quality. It may require some homework, but for every defined market, regardless of price point, there is an applicable spathiphyllum.

Popular spathiphyllum cultivars

Cultivar	Pot size	Source	Production area
Cupido	3 to 6 in.	Seed	Europe
Quatro	6 in.	Seed	Europe
Petite	3 to 8 in.	Tissue culture	Worldwide
Starlight	4 to 10 in.	Tissue culture	U.S.
Viscount	6 to 10 in.	Tissue culture	Worldwide
Gigant	6 to 10 in.	Tissue culture	Europe
Supreme	8 to 14 in.	Tissue culture	U.S.
Lynise	6 to 10 in.	Tissue culture	U.S.
Sensation	8 to 17 in.	Tissue culture	Worldwide

Gary R. Hennen is president and Steven E. Hotchkiss is director of marketing and sales, Oglesby Plant Laboratories Inc., Altha, Florida. December 1995.

Managing Phytophthora on Spathiphyllum

A. R. Chase

Phytophthora aerial blight on *Spathiphyllum* spp. (caused by *P. parasitica*) has been increasing in frequency and severity. Symptoms are large black or brown dead spots up to 1 in. wide on leaf edges or centers. Spots are wet and mushy under moist conditions but dry if foliage is kept dry. The disease is especially common and severe on large spathiphyllum grown in shadehouses, as conditions for incidence and spread are ideal much of the year.

Controls include keeping foliage dry, growing in a sterilized potting media, and using raised benches away from the native soil when possible. Soil drenches with a labeled fungicide are generally effective. Using fungicides through the overhead irrigation system is a tempting way to treat extensive areas of large spathiphyllum, but you must be sure to check labels for legal application methods. This type of application may not be effective if dosage rates and amount of irrigation time are not accurate. There are a number of other serious diseases of spathiphyllum that can mimic symptoms of phytophthora aerial blight, including erwinia and myrothecium leaf spots. An accurate diagnosis is the most important tool for controlling any disease.

A. R. Chase, Chase Research Gardens, Phoenix, Arizona. April 1996.

Sunflower (Helianthus)

Success with Pot Sunflowers

Brian Whipker and Shravan Dasoju

Pot sunflowers have been popular in Europe for a number of years. Popularity in the North America market is still increasing, and with the introduction of new improved cultivars such as Pacino and Sundance Kid, it's likely to expand. Pot sunflowers are a quick crop to produce, and they offer an opportunity for growers to capitalize on the current consumer craze for the plant.

Varieties

Big Smile, Pacino, and Sundance Kid are the main pot sunflower varieties produced. Big Smile has been available for a number of years. Four- to 6-in. flowers have yellow petals surrounding a black center. Plants produce an individual flower, and it's common for three plants to be grown in a 6-in. pot. Pacino is a new cultivar introduced by Benary Seed Co. in 1996. Flower petals and centers are yellow, and plants are prolific pollen producers. Plants have sturdy stems with a single, dominant 4- to 5-in. flower that blooms first, followed by four to six additional secondary flowers that open four to six days later. Plant size is sufficient for one plant per 6-in. pot. Sundance Kid is a recent release from Sakata Seed Co. The bronze- to pure-yellow-shaded flowers are 4 to 6 in. across, and multiple flowers are possible, although in trials at Iowa State University, most plants only produced a single flower. Sakata recommends three plants per 6-in. pot.

Propagation

Pot sunflowers are propagated from seed. Seed counts are 810 (Big Smile), 900 (Sundance Kid), and 1600 (Pacino) seeds per oz. Germination occurs in two to five days when germinated at 70° to 75° F (21° to 24° C). Seeds can be sown directly into 1203 or 1204 flats and transplanted in two to three weeks at the two- to four-leaf stage into the final container. Some growers direct sow seeds into final containers.

Culture

Plants should be grown with 72° F (22° C) days and 65° F (18° C) nights. Warmer temperatures will cause plants to stretch. The number of days from sowing until flower varies by cultivar. Big Smile flowers in the shortest amount of time, seven to eight weeks. This short production time offers the ability to produce a quick cropping schedule. Sundance Kid is an eight- to nine-week variety. Pacino has the longest production time at nine to eleven weeks. Crop time varies with season. Pacino plants bloomed in eleven weeks when sown in January or February and nine to ten weeks when sown in summer or early fall.

Most sunflower varieties are day neutral but will flower more quickly under short

days. This is especially true for Big Smile and Sundance Kid. Under short days, plants can be excessively short, and less leaf canopy is produced. Provide supplemental lighting during winter months to improve plant quality.

Nutrition

Recommended clear liquid feed fertilization rate is 150 ppm nitrogen. Less than 75% of nitrogen should be in the nitrate form. An experiment at Iowa State University found that in order to match the total shoot nutrient content of Pacino pot sunflowers, the optimal fertilizer ratio of N:P:K:Ca:Mg was 8:1:10:4:2. This would result in a recommended rate of less than 15 ppm phosphorus, 180 to 220 ppm potassium, 100 ppm calcium, and 50 ppm magnesium for optimal growth. Pot sunflowers are heavy potassium feeders, which would be expected because of potassium's essential role in maintaining stalk strength. Foliar tissue standards for pot sunflowers are listed in the table.

Diseases

The most serious disease of pot sunflowers is pythium root rot. Pythium is usually present in most media. As the first line of defense, avoid growing conditions that stress the plant: continuously waterlogged media, cool media temperatures, or cool air temperatures. If necessary, apply monthly drenches of Subdue or Banrot as a preventive.

Botrytis can also be a problem on foliage and flowers. The vegetative growing tip of Pacino forms a tight whorl before the flower bud is visible. This whorl is susceptible to water droplet retention and botrytis infection. Avoid overhead watering, and control moisture condensation within double poly structures to prevent the problem.

Other potential root rot diseases include phytophthora and verticillium wilt. Sunflowers are also susceptible to a number of foliar diseases, the most common being powdery mildew, downy mildew, alternaria blight and leaf spot, rust, and septoria leaf spot. Control foliar diseases with preventive fungicide applications if needed. Aster yellows, which is transmitted by aster leafhopper and sunflower mosaic and is caused by cucumber mosaic virus (CMV), has been reported on sunflowers grown as a seed crop.

Insects

Thrips is the most common insect pest. They feed on leaves, flower petals, and pollen. Tame and Orthene can be used for control. Ideally, keep thrips under control before flowering to avoid the potential of flower phytotoxicity. Whiteflies and a number of caterpillar species can occasionally be a problem.

Marketing

An informal survey of growers reveals that most are growing Big Smile and Sundance Kid with three plants per 6-in. pot and Pacino with either one or three plants per 6-in. pot. Wholesale prices ranged from $1.95 to $3.05 per 6-in. pot, and retail prices were $3.50 to $4.50 per 6-in. pot.

Foliar tissue standards

Nutrient	Recommended concentration
Nitrogen (%)	5.0 to 6.0
Phosphorous (%)	0.70 to 0.80
Potassium (%)	5.4 to 6.3
Calcium (%)	2.2 to 2.5
Magnesium (%)	0.59 to 0.80
Boron (ppm)	43 to 53
Copper (ppm)	6.7 to 7.2
Manganese (ppm)	67 to 99
Molybdenum (ppm)	0.42 to 1.8
Zinc (ppm)	77 to 115

Note: Values are reported on for dry-weights based on a limited number of plants. Tissue samples were collected at bloom from Pacino pot sunflowers grown with 100 and 200 ppm nitrogen.

Brian Whipker, floriculture extension specialist, and Shravan Dasoju, graduate research assistant, Department of Horticulture, Iowa State University, Ames, Iowa. May 1997.

※

Pot Sunflower Culture Tips

Brian Whipker and Shravan Dasoju

The ongoing consumer craze for sunflowers makes pot sunflowers an ideal crop. New research on height control and postharvest care from Iowa State University can help you produce better quality pot sunflowers.

Height Control

When pot sunflowers are adapted for greenhouse container production, they can be disproportionately large relative to the pot size because of growth habit, making them less attractive to consumers. Growth regulators solve the problem. For height control on the variety Pacino, for example, Benary recommends applying multiple foliar sprays of B-Nine at 1,500 ppm, starting three weeks after sowing. Repeat applications as needed.

Researchers at Iowa State University studied the effectiveness of plant growth regulator foliar spray treatments and drench rates as chemical height control for pot sunflowers. The foliar spray experiment on Pacino pot sunflowers used single applications of B-Nine at concentrations from 1,000 to 16,000 ppm; Bonzi at 5 to 80 ppm; and Sumagic at 2 to 32 ppm. Based on these concentrations, market-sized plants grown in 6-in. pots were produced with Sumagic concentrations between 16 and 32 ppm or with B-Nine concentrations between 4,000 and 8,000 ppm. At the concentrations used, Bonzi foliar sprays had little effect on height control.

We also conducted growth regulator drench experiments. Bonzi drenches of 2, 4, 8, 16, or 32 mg active ingredient (a.i.)/pot were applied to pot sunflowers. All Bonzi concentrations applied reduced plant height by about 27% when compared to the untreated control, but 16 and 32 mg a.i./pot produced excessively short plants. Plant diameter was also decreased by about 16% at 2 and 4 mg a.i./pot of Bonzi when compared to the untreated control. Bonzi concentrations had no effect on days from potting to flowering. Drench concentrations of 2 to 4 mg a.i./pot of Bonzi produced optimum height control for to 6-in. pots. Based on observations, a drench rate between 2 and 8 mg a.i./pot of Bonzi should be suitable for a 4- or 4½-in. pot. Use lower rates during winter.

Postharvest

Another research project at Iowa State University investigated the influence of fertility level on pot sunflower growth and postharvest quality in interior conditions. Pacino plants were fertigated on ebb-and-flood benches with 100 or 200 ppm nitrogen. Fertilization rates were held constant from potting until day forty-five; then the fertilization rates were continued, decreased, or stopped on day forty-five and day fifty-five, giving a combination of nine fertilization subtreatments. At bloom, plants of each fertilizer treatment were moved into interior conditions with artificial lighting and were evaluated five, ten, and fifteen days after moving for postharvest quality. Fertilizer treatments had no effect on number of days to flower, plant height, or flower diameter. Plants fertilized with 100 ppm nitrogen from potting until day forty-five in combination with stopping fertilization on day fifty-five had better plant grades than plants grown with a constant 200 ppm nitrogen. Plants fertigated with 100 ppm nitrogen also had a postharvest life of eleven to twelve days compared to nine days for the constant 200 ppm nitrogen fertilization rate.

For extended shelf life, fertilize pot sunflowers grown on ebb-and-flow benches with 100 ppm nitrogen, and stop fertilization seven to 10 days before bloom. A rate of 150 ppm nitrogen is recommended for top-irrigated plants.

Brian Whipker, floriculture extension specialist, and Shravan Dasoju, graduate research assistant, Department of Horticulture, Iowa State University, Ames, Iowa. June 1997.

Trachelium

Trachelium Production Tricks

Trachelium caeruleum, or throatwort, has become a popular cut flower that's used as filler in bouquets. Its panicle flowers come in striking lavender and dark blue as well as creamy white and pastel pink. Most growers purchase plugs rather than propagate this tender perennial. Here are some cultural tips from Vegmo Plant, Rijsenhout, the Netherlands.

Trachelium is easily started from plug seedlings and may be greenhouse-grown in containers or field-grown in ground beds.

Trachelium seed germinates in five to seven days at 65° to 68° F (18° to 20° C). Leave seed exposed to light for germination. Keep seed and seedling trays under short days to prevent premature flowering.

Crop time varies from ten weeks for July plantings to seventeen weeks for November through January plantings. Plant out at a density of fifty-four plants per sq. yd. for single-stem production, eighteen plants per sq. yd. for pinched production.

Lighting

Long days encourage flowering; short days inhibit it. Provide continuous sixteen-hour days, or use night interruption lighting (mum lighting). Light at an intensity of 15 watts per sq. m. Light until harvest to avoid misshapen, pyramidal flowers.

HID lights will make plants flower faster; incandescent lighting will promote faster growth.

For crops planted from the end of May until early July, encourage vegetative growth before initiating flowers by applying short days (12 hours) during weeks two and four. The exact length of short-day treatment depends on light levels and temperatures.

Temperature

Trachelium is native to the Mediterranean and prefers cooler temperatures. Therefore, it doesn't do well outdoors in the hot summers of the South or Midwest. For fall and winter crops, grow at 50° to 52° F (10° to 11° C) for the first three weeks, then raise night temperatures to 58° F (14° C) and days to 60° F (16° C). Spring crops can be grown at 60° F (16° C) nights and 62° to 68° F (17° to 20° C) days. In colder climates, additional heat may be necessary to maintain 52° F (11° C) nights and 62° F (17° C) days. Note: While warmer temperatures will speed up flowering, they may also cause poor flower quality and thin stems.

Fertilization

Plants require average fertility. During bud formation, apply additional potassium nitrate. Be sure to rinse off plants afterward to avoid burning. Don't use ammonium fertilizers on trachelium.

Also, trachelium is very sensitive to high manganese levels in the soil. Use care when steaming, as steam releases manganese bound in the soil, which can result in manganese toxicity. Toxicity symptoms will show as leaf damage during the last stage of development. Steaming for four hours or less and using iron chelate prevents leaf damage. Apply iron chelate three to five weeks after planting.

Postharvest

Harvest flowers when one-quarter to one-third of panicles are three-quarters open. Flowers normally have a ten-day to two-week vase life in water. Silver thiosulfate can improve keeping quality. Store stems in water for twenty-four hours at 39° F (4° C) before shipping.

Excerpted from the 16th edition of the **Ball RedBook.** *September 1997.*

Trachelium caeruleum

Ir. Jeroen Ravensbergen

In early days, trachelium species were used to treat tracheal diseases. Today *Trachelium caeruleum* is grown as a cut flower valued for its long stem and a 3- to 6-in. umbel in blue, purple, white, or pink. In areas with low-light winters, trachelium can be produced from spring through fall. Growers in areas with higher winter light can produce trachelium year-round. While many improvements in existing varieties have been made, the search for new trachelium varieties continues. With time, breeders will introduce new varieties with greater uniformity, better low-light tolerance for year-round flowering in Northern climates, and a broader spectrum of colors.

Culture

Timing

When planted from November through January, trachelium will flower in approximately fourteen to sixteen weeks. Plantings from February and March flower after twelve to fourteen weeks. Spring plantings require ten to twelve weeks, and summer plantings require ten weeks. Plantings made at the season's end can be harvested for an additional six weeks by delaying bud initiation and flowering to extend the harvest period. At harvest, take the entire central leader.

Planting density

Single-stemmed trachelium can be grown as close as six plants per sq. ft. To produce pinched (branched) trachelium, plant only two plants per sq. ft.

Temperature

Maintain temperatures at 52° to 55° F (11° to 13° C) during the first three weeks of autumn and winter plantings, followed by 55° F (13° C) average nighttemperatures and 60° F (16° C) days. If light intensity is high from February to March, maintain night temperatures of 60° F (16° C) and day temperatures of 65° F (18° C). In late spring and summer, 60° to 62° F (16° to17° C) night temperatures are ideal, with a maximum of 78° F (26° C) during the day.

Watering

Trachelium caeruleum requires uniform moisture. Irrigate from above, beginning at planting time until plants are about 12 in. tall, then water at soil level.

Fertilization

Trachelium caeruleum requires an average fertilization level. Provide additional potassium nitrate when buds begin to form. If a liquid feed is used, always rinse plants after fertilizing to prevent burning. Trachelium is easily damaged by excess manganese in the soil, so don't steam-disinfect soil, because this releases soil-bound manganese, which can damage leaves in the last stages of development. It is possible to prevent some of the damage by providing extra iron.

Shading

When light levels are high, whitewash in the late spring and summer to produce long flower stems.

Photoperiodic lighting

Trachelium crops planted from November through March require additional lighting. Summer plantings need light after August 20. Increase the daylight period to about sixteen hours, or provide cyclical light throughout the night.

To produce long stems on late May and June plantings, provide twelve hours of darkness for two to three weeks beginning the second week after planting. Since trachelium is a quantitative short day plant, its reaction to short days is influenced by cumulative light intensity as well as day length. During cloudy, cool weather, bud initiation is slowed, and the need to apply long nights is less.

Disease

Treat the soil for *Rhizoctonia solani* before planting trachelium. Pythium can also be a problem when plants are small. Preventive spraying for disease and insects is required during cultivation. Be aware of sclerotinia during the growing period when relative humidity is high.

Varieties

Blue varieties include Wonder, Blue Umbrella, Merii Blue (summer production), Lake Success (fall production), and Moonshine (dark blue). Lake Superior is purple, Lake Avalon is a rose-pink variety, and Lake Powell and White Umbrella have white flowers.

To expand the existing varieties, Vegmo Plant will introduce two new varieties in 1995. Summer Blue Wonder has blue flowers and is recommended for early summer plantings. Summer Lake Superior, also for summer, is an improvement of Lake Superior.

For Southern climates, choose varieties that are well suited to summer production such as Merii Blue and Summer Lake Superior. Lake Success can be grown for winter production in the South.

Ir. Jeroen Ravensbergen is a cut flower researcher for PanAmerican Seed Europe B.V. December 1994.